Pioneer Days
in the Southwest
from 1850 to 1879:
Thrilling Descriptions of
Buffalo Hunting, Indian Fighting
and Massacres, Cowboy Life
and Home Building
(1909)

by

Charles Goodnight

Originally published
1909

Introduction

Ever since the Pilgrim Fathers landed on the New England coast, and a little later at Jamestown, Va., surrounded as they were by primeval forests, inhabited by savages, and savage animal life, every step of progress and the introduction of civilization was beset by perils and hardships of such magnitude that it has provided a wide field for the historian and a yet wider field for the writer of romance. These conditions of menacing perils, privations and hardships developed a race of people of such indomitable courage and determination, and yet of such simplicity of character, hospitality and kindness of heart that the world became better because they lived in it. Step by step these hardy pioneers subdued the wilderness and made it blossom as the rose. Our imagination has been fired by such names as Boone, Kenton and the Wetzels of the dark and bloody days in Kentucky, and later farther west on the great plains and the Rocky Mountains we have other historical names, Kit Carson, Buffalo Bill (Cody), Payne and others, but very little has ever been written about the great southwest, where the red men of the prairie made their last struggle for supremacy, and where they fought out their last undying hatred against the innocent settlers and pioneers, who with all they held dear on earth, hewed out homes for themselves and the coming generations amid the most indescribable dangers from their savage foes. Pioneer Days is written by the rank and file who were the true heroes and heroines, who suffered and gave their lives and the lives of those near and dear to them, in order to lay the foundation for future happy homes, peace and prosperity. The writers of this book are the small remnant yet left who were the actual participators in these early struggles, and they give their experiences, unadorned, without any claim to literary merit; for the writers are now old, and most of those who shared those dangers and hardships have already passed over the river. When you read their simple statements of facts of Indian massacres, of terrible suffering and privations, so unassumingly told by them, I ask you, who have had the advantage of schools and Christianity, and refinement, of which they were almost entirely deprived, to cover their rough and often ungrammatical sentences with the cloak of Christian charity, and interline them with garlands of flowers, romance and chivalry which truly belongs to them.

With this short preface we give you not fiction or idle tales, but unadorned truths and conditions that fortunately have passed out forever.

Hoping our simply told reminiscences of the early days will meet with your approbation, and that you will treasure our memories in love, for very soon indeed, the burdens and the joys of life will be no more for any of us, and we can only hope to live in your appreciative memories.

NOTE BY THE WRITER, E. DUBBS: The statement that Indians would not kill an old person, a cripple or a deformed person, may have been correct in rare instances, but the statement is incorrect as a general proposition. My own observation, and that of hundreds of others, has been that when on the war-path, they spared no one. They would, and did to my own personal knowledge, take the babe from its mother's arms and beat out its brains against the door frame, and then work their pleasure on the mother and cruelly mutilate and kill her afterwards, sparing no one, young or old, male or female.

Contents

PIONEER DAYS IN THE SOUTHWEST

CHAPTER I. CHARLES GOODNIGHT. BY EMANUEL DUBBS.

No history of pioneer days would be complete without the name of Charles Goodnight. While Mr. Goodnight has a state and national reputation, the people of the Panhandle of Texas feel that they are especially honored in owning him as a citizen, and he and his estimable wife had, and now hold a place in the hearts of oldtimers as well as later settlers, that would cause the people to condemn any writer who failed to give to them that mete of praise which they so richly deserve, and place their name at the head of the highly honored galaxy of heroes who contended with and finally overcame every obstacle and danger of a country entirely given over to savagery at the time of their advent.

Charles Goodnight is perhaps more extensively known than any other western ranchman, cattle owner, and pioneer, or any man of the present day. A man who has done more to help to develop this great Panhandle country and its diversified industries than probably any other, except T. S. Bugbee, who was the second settler in the Panhandle, while Goodnight was the first. I will refer to this again in its proper place.

The writer of this sketch speaks from personal knowledge, for he has known Mr. Goodnight more or less intimately for thirty-three years. In the fall of 1875 and 1876 Mr. Goodnight came to our buffalo and meat drying camp, on Kelly creek, preparing for his permanent location in Paloduro Canyon. All the other characters represented in this book are their own contributors and tell their life's story in their own way, and all of them simply and modestly. But it was different with Mr. Goodnight. I went to see him personally at his beautiful home, and found him very busy getting ready to start on a trip to the Pacific coast with his wife and two nieces, the Misses Dyer. Yet busy as he was, he accorded me every attention possible. Mr. Henry Taylor of Clarendon, than whom a truer friend and gentleman never lived, accompanied me, and we were both entertained royally, and most hospitably. And a lively scene burst upon our view; four buffalo cows were already hanging up, beautifully dressed on their separate cranes, their beautiful black furry robe hides carefully salted and rolled up ready to be shipped to their purchasers back east.

"Mr. Goodnight, what do you get for these hides?" was my question.

"They average me $100 each," was his answer.

Shades of Moses! Think about it. And it does not seem so very long ago when I was glad to get as the top price $2.50 each. Upon further inquiry he informed us that each buffalo killed that day netted him about $350. I informed him of the object of my visit. His answer was, "I haven't time to write what you want, but after dinner when I dispose of some of these pressing matters I will give you such data as you want," but he says, "I want it understood I only do this to help you out and as a personal favor to yourself."

With this introduction, dear readers, I will give just as near as I can in his own words a short life sketch of Mr. Goodnight, for his modesty prevented him from giving himself credit in times of trial and danger, when nothing but supreme nerve, coolness and determination carried him safely through and preserved his life.

"I was born in Macoupin county, Illinois, March 5, 1836. I am now in my seventy-third year. After the death of my father, in 1841, my mother married again, and moved to Texas, and first settled in Milam county, in 1846, when I was in my tenth year. In 1857 I went to Palo Pinto county, then on the extreme frontier. At that time all of northwestern Texas, lying beyond the east fork of the Trinity, was comparatively a wilderness and its scattered settlements were continuously in danger of Indian raids. Dallas was a mere village, its first house having been built five years before, while the military post, known as Fort Worth, was not established until 1849. The country adjacent, now a solid farm, was then covered with a rank growth of mesquite grass, affording pasturage for all kinds of large game common to the latitude. Buffalo herds roamed all of this region, and I first encountered them on the west side of the Trinity river, within two miles of Dallas. I attended public school in my early childhood, but our removal to Texas prevented me from securing an education other than that acquired in active everyday life. Until my nineteenth year I was engaged in farming, a life which in all new countries is one of ceaseless hardships and toil, but doubly so in a region so far removed from civilization as Palo Pinto county was then. All this western border was then hampered by the harrassing bands of savage enemies, and with the consequent need of incessant watchfulness against their raids, the settlers enjoyed but few respites from privations and dangers.

"In 1856 I started on an overland trip to California, accompanied by a young man about my own age, J. W. Sheek by name. Our outfit consisted of a bull team and a wagon and three horses. When we reached San Saba river we concluded that Texas was large enough to supply an ample field for our energy, so we turned back over the same route just traveled. At the crossing on the Brazos we encountered a cattle man who proposed letting us have a herd to handle on shares. There were 430 head in the bunch, principally cows. The contract was: We could hold them where we thought best, and we could brand for ourselves one-fourth of the increase. We moved the cattle back to Palo Pinto county and located in Keechi valley. As the end of the first year's branding resulted in only thirty-two calves for our share, and as the value was about three dollars per head, we figured out that we had made between us, not counting expenses, ninety-six dollars. It was a gloomy outlook at the time. However, we determined to hold to our contract, and the herd intrusted to us became one of the largest and finest in the country, at the expiration of our contract we had 4,000 head for our share.

"During the last years of the civil war, in 1865, I served as scout and guide for a frontier regiment, which was guarding the settlements from Indian attacks, and my cattle interests were in consequence neglected, and I suffered great losses. The confederate authorities had taken many ot them without paying a cent, Indians had raided our herds and cattle thieves were branding them, to their own benefit without regard to our rights."

The writer wishes to note a fact here that no doubt the reader has discovered that during all this time Mr. Goodnight says nothing about his own encounters with the Indians and outlaws, and when I questioned him about it he said: "I never talk about it. I leave that to Tom Pollard or Jim Tackett, who were with me and can give you stories of Indian fighting that I have now forgotten."

PIONEER DAYS IN THE SOUTHWEST

Mr. Goodnight continued: "Upon my return from the army I became disgusted with the then existing state of affairs, and accompanied by Col. Dick Carter, C. C. Slaughter, Dick Jorvell and George Lemly, we started to Mexico in search of a more favorable business location, but the party never reached its destination. While making our way through thickets on the Devil's river, Slaughter was painfully injured by the accidental discharge of Lemly's gun, and we turned back to a point where we could obtain medical attendance. Upon my return to Palo Pinto county I found my affairs in a worse shape than ever, and determined to carry what cattle I had left elsewhere. I now turned my attention to New Mexico as a field offering fair advantages, and arranged to drive there with a Mr. Mosely, who was the owner of a very large herd. But the dangers of the trip proved too great for Mr. Mosely, who changed his plans at the last minute, going to Alabama instead.

"At this juncture I met Oliver Loving, who readily fell in with the idea to take our joint herds to New Mexico, or even to Colorado, if we found nothing to suit us in the first territory. The start was accordingly made, the outfit consisting of myself, Loving and fourteen men. We drove the cattle by way of Fort Belknap, to Fort Sumner, New Mexico, opening, what has since become known as the Goodnight trail. We found purchasers for the cattle at Port Sumner, and the entire herd was sold for eight cents per pound. This was an incentive to us and we determined to return and buy another herd and rush them through the same season. Turning back on the trail, with the proceeds of our sale loaded on our pack mules, an incident occurred which came very nearly wrecking our new found hopes. While in the roughest and most dangerous part of our journey back, and when we were using every precaution to prevent giving notice to the red skins of our presence, a mule bearing six thousand dollars in gold in the pack, broke away in the darkness, scattering the provisions she carried in every direction. The gold, however, was saved, for I grabbed a rope that was dangling from the pack and checked the frantic animal after being dragged quite a distance down a rocky slope at the risk of a broken neck. The gold was saved, but our provisions were entirely lost, and there was no way to get them renewed. We traveled for eighty-six miles without anything to eat until we got to the Pecos river, where we accidentally ran across a man who divided his meager stock of food with us. We continued our return without further incident, and a second herd was secured and driven through that same season. We accomplished this trip in thirty-three days, and sold them out again to good advantage. On our return this time we concluded to avoid the pack mule accident, so we bought a wagon. Most of our money was in silver and it took a wagon to haul it. Among the silver we had a great many dimes and small change, which added greatly to the bulk and was an undesirable load to carry, and I felt greatly relieved when the 'nigger' cook said, 'Marse Goodnight, I takes the very smallest money dere is." When he had his hat piled full of dimes and quarters he said, "Tse gwine tei make dem town niggers open der eyes powerful." The next year Loving and I started on our third drive, but our fortunes changed from the very start. While we halted for the night near Camp Cooper, the Indians attacked us, shooting one man in the head with an arrow, but not fatally. I had here a narrow escape. Very tired by our day's hard ride, I was sleeping on a buffalo robe by the fire and an arrow sent with all the force of a strong bow, struck the edge of the robe, deflecting sufficiently to pass under me, barely missing my body."

(Note by the writer.)

I desire to say to those of my readers who have not seen an Indian handle a bow and arrow, that they are a powerful weapon at close range. I had an Indian show me what he could do while in my buffalo camp on Oaks creek in the Panhandle. A flint dry buffalo hide was leaning up against the door of our tent. He fitted an arrow to his bow, drew it back to the head, and let drive, and it penetrated the hide clear up to the feather, while two thickness of dry buffalo hides were bullet proof. Even the "big fifty" Sharp's rifle could not penetrate them.

Mr. Goodnight continued: "As I sprang up I heard the leader of the Indians shout out something which I took to be an order to stampede the horses, and running out quickly I reached the herd ahead of them and the first Indian I saw coming I fired at him, and believing, we supposed, that the horse herd was under a heavy guard, the attacking party retreated. Next day we recovered all the horse stock but one mule, but over three hundred cattle were driven away. Some days after this, while on Pecos river, Mr. Loving left the outfit with a single companion, intending to press on ahead of the outfit to Fort Sumner and secure an Indian contract by the time of our arrival with the herd. It was a dangerous undertaking. I warned him to use the greatest precaution to avoid the Indians and only travel by night and hide through the day time. In his impatience Mr. Loving neglected these precautions and was attacked by 600 Indians at a point near the present site of Eddy, New Mexico. Wounded in the wrist and side, he and his companions were forced to seek shelter under the bank of the river, where the savages were too cautious to press them. The Indians took their position on either side of the river, and posted one in the stream down the river to prevent escape by swimming. Mr. Loving seeing that there was no chance to pass them in his wounded state, he told his companion, J. M. Wilson, to take his gun, as it was the best of the two, and try to reach me with information of his situation. Wilson made three efforts to swim with his gun and clothes on, but failed each time. At last, he pulled off his clothes and concealed them with his arms, again trusting himself to the current, staying under the water as long as he could, passing the Indian lines. Three days afterwards he reached our camp, which was close to Adobe Walls, naked, barefooted, weak and exhausted, and so changed in appearance that his own brother, who was in camp, failed at first to recognize him. I immediately started with six men, one of whom was Mr. Campbell, who was traveling with the outfit for safety, and who was afterwards a prominent banker at Comanche, Texas. Wilson had described the spot where Loving had been left, so accurately that we were certain of finding it, but I had no hope of finding him alive. In course of time we reached the spot where the first fight had occurred and where the two men had left the road, and where they fled through a narrow strip of brushy bottom to the river bank, and here we found the trail of fully 600 Indians in pursuit. We all believed that Loving had been killed, but if not, we determined to rescue him or die in the attempt. The Indians were gone and no trace of the wounded man was found, though we found the spot where he had made his stand and could see that the Indians had tunneled within six feet of his position. We also found Wilson's gun and clothing where he had hidden them. I afterwards learned with an accidental meeting with Jim Burleson, another 'old-timer' that Mr. Loving had evaded the Indians by going several miles up the river instead of down, where he was found by some Mexicans and taken to Fort Sumner, but the five days' exposure and suffering before he was found proved too much, even for his iron constitution, and he died shortly after reaching the fort. I changed my course and drove on through to Colorado with the herds, but the losses on the route so reduced them in numbers that I barely realized from the sales first cost.

"I continued driving cattle to Mexico and Colorado markets for three years longer, and then found myself in possession of $72,000, a part of which belonged to the Loving estate, and was turned over to the heirs of my old partner. The last year spent on the trail I worked in conjunction with John Chisum, clearing a net profit of $17,000. The year 1871 was a fortunate year for me in more than a financial way, for I met and married a most estimable lady whose counsel and influence has ever since been to me a guiding light for good deeds, and godly living. Her maiden name was Mary Ann Dyer. Though born in Tennessee, she had been principally reared in Texas. Her father, Joel Henry Dyer, a prominent lawyer and ex-attorney general of Tennessee, emigrated to Texas when the country was yet young, settling in Fort Worth. He was a resident of Belknap for a good many years and Mrs. Goodnight was sent to the best schools in northern Texas, and afterward taking up the profession of teaching." The writer of this incomplete life sketch believes this a fitting and appropriate time to say a few words regarding Mrs. Goodnight from personal knowledge. Her wide Christian philanthropy is not only felt at home among home people, but has reached out all over the state. There are many boys and girls scattered all over this broad domain who, now grown into manhood and motherhood, who call her name blessed. She, aided by her noble husband, built entirely with their own means the original Goodnight college and endowed it with the land (the number of acres I do not know), and for years kept up an excellent faculty almost entirely at their own expense.

Mrs. Goodnight was a mother to all the boys and girls intrusted to their care. Her true worth, her liberality, her charity, her consecrated Christian life, can only reap its full fruition in the home of "many mansions."

PIONEER DAYS IN THE SOUTHWEST

Mr. Goodnight says: "After I married I thought I would no longer follow my wild trail life, as I had ample means. I concluded to settle down and take up ranching instead. I purchased in 1868 a large tract of land in Colorado near Pueblo, and in 1871 I bought, forty miles from Trinidad, a large ranch, where I lived for a time and temporarily enjoyed the reputation of being the largest farmer in eastern Colorado. When I first interested myself in farming here, corn commanded a ready sale at six and seven cents per pound. But the building of railroads soon put an end to these fancy prices, and they shipped in all kinds of farm products cheaper than we could raise them. I now assisted in organizing the Stock Growers' Bank of Pueblo, and began investing extensively in different enterprises, but in the panic of '73 I saw my property to the amount of $100,000 suddenly swept away, leaving me nothing in the way of working capital, save a herd of 1,800 head of cattle. With this small remnant of my former capital I determined to again return to the pursuits of my earlier life, fully determined if possible to retrieve my losses in the same manner that my first start in life had been gained. I had a very good knowledge of the entire portion of northwest Texas. I believed that this section, considering its location as to markets, and practically opened up for settlement by that fearless body of men, the buffalo hunters, who by killing out the buffalo stopped forever the terror of the settlers and the cattlemen, the depredating tribes of plains Indians who had for centuries roamed and fought over this fair and productive part of Texas, known as the Texas Panhandle.

"In 1876 I moved my herd into the Panhandle, following an old Indian trail, which led into the banks between canyon Blanco and the main canyon Paloduro, where we came to the cap rock over which the trail led, while probably the most practicable place to enter it for many miles around, but the descent was so abrupt and steep that we were compelled to take our wagons to pieces, first unloading them, and let them down into the valley below with ropes. We had to do the same thing with their contents. We had heavy loaded teams of provisions and many other necessary things for establishing a permanent camp or quarters. I located in Paloduro canyon with my herd of 1,800 head. At that time the nearest settlement eastward from us was Henrietta, 200 miles northeast, seventy-five miles to Fort Elliott, which had just been established the year before, having then only a small nucleus of settlers. North eighty miles was the Canadian river, where another cattleman located his family, T. S. Bugbee. My wife's nearest neighbor was Mrs. Bugbee, eighty miles away. Think of it, my lady readers, the lonesomeness of those two women, both by nature and education qualified to adorn the most exclusive society, but with willing hands and hearts helping their husbands to lay the foundation for their fortunes, and preparing the way for the future development of a grand civilization. In the fall of 1877 a small settlement was started in Donnelly county known as old Clarendon, about five miles north of the present little and beautiful city of the same name, on the Fort Worth and Denver railroad.

"In 1877 I formed a partnership with Lord John G. Adair of Weathdair, Ireland, under the firm name of Adair & Goodnight. Mr. Adair invested $372,000 in the business, of which I had the management, I owning one-third interest. A large tract of land was purchased, none of which was surveyed. The pioneer surveyors, Gunter and Munson, under the authority of the state government, ran out the lines, assisted by our own townsman (now dead), T. S. McClelland. The land purchased by the firm included the Paloduro ranch, originally located by me, and now selected and occupied as 'headquarters' and occupied as our home until 1879. The buildings erected by me are yet standing and became the headquarters of the J. A. outfit."

There are some incidents of interest of which the writer is cognizant, and of which Mr. Goodnight in his modesty did not speak in my interview with him. Incidents that were factors in putting down lawlessness and cattle stealing in its different forms. Also showing Mr. Goodnight's determination to assist as far as was in his power the weak authority of the courts which were yet very crude and had not as yet laid off their swaddling clothes.

In my own reminiscences, I stated that Wheeler county organized and elected county officers in the spring of '79, her courts having jurisdiction over the vast territory of twenty-six counties which in the last year had an influx of all sorts of cattlemen, sheepmen, a few settlers, and many "rustlers." On the whole, the country was blessed in her sheriff, Mr. Henry Fleming. Mr. Fleming was a type that now no longer exists, or if they did, could no longer be elected to such a responsible office. A saloon man and a gambler by profession, yet in the discharge of his duty as an officer, firm and brave, and in the language of Mr. Goodnight: "The only man who could preserve the law, arrest men in their wild drunken orgies without bloodshed, and for the four years that he was sheriff he made many arrests, but never shot a man or was shot." However diligent, it was impossible to keep down lawlessness. About three weeks before our first term of district court, Mr. Goodnight came to me and said that there was a great deal of cattle stealing going on all over the Panhandle, that the officers of the law could not, except without outside assistance, ferret out and bring to trial the lawbreaker over such a vast territory. And he proposed that we prepare a proclamation, asking all stockmen, large and small, to meet in Mobeetie, the county seat of Wheeler county, on a certain day during district court week, in order to formulate plans to assist the officers of the law to break down lawlessness. This was done, and in that meeting was born the Panhandle Stock association, with Charles Goodnight the first president.

The town of Clarendon and small settlements surrounding it in Donnelly county, of which we spoke before, was composed of such men as Mr. Carhart, H. B. White (the latter judge of the county until he died), T. S. McClelland, Mr. Stanton, and other citizens, determined not to tolerate saloons, whiskey, and gambling dives. But Mr. McCarney at Mobeetie sent one of his men from their town with teams and wagons loaded with all things required to open up a first class frontier institution. A combination of sin, drinks, gambling, dance hall and lewd women, and armed with a government license for selling intoxicants. The people of Clarendon were very much opposed to it, but did not know how to prevent the man from opening his dive. Fortunately Mr. Goodnight just came to town with a number of his men from a round up. He did not want a saloon, either, to demoralize his men, so he told the people he guessed they need not have a whiskey dive unless they wanted one, and as they declared they did not want anything of that kind Mr. Goodnight went down to where the gentleman was camped and told him to hitch up and pull out. Both were armed, and it looked for a little while which one would get his gun the quickest. The man, looking into Mr. Goodnight's eyes, concluded he would turn around and go back, and as he expressed it afterward, "It did not look healthy around there."

Now dear reader, after giving these side lights on my own responsibility, I continue Mr. Goodnight's narrative:

"The partnership between myself and Adair continued until 1888. At one time we had 100,000 head of cattle under our management and to care for them over such an extensive range we employed a little army of men called cowboys. Everything was conducted on a very extensive scale. At the time of the dissolution of the partnership there were 63,000 head of cattle on the ranch. In the division of the property I kept what was known as the Quitequa ranch. I now moved sixteen miles north of the J. A. ranch to what is now known as Goodnight station on the Fort Worth and Denver railroad in Armstrong county, where I built and occupied my present home and have lived here ever since." This closes his personal narrative.

The reader of this short life sketch, as personally related to me, has no doubt wondered why it is so short and barren of details. The real cause of this is that Mr. Goodnight is extremely modest and does not desire to pose before the public as a hero. I feel myself fortunate in having obtained from him as much data as I did, and I believe that he was free with me, more perhaps, because I am an old-timer, and of our pleasant personal associations of the past.

The Goodnight home, park and ranch has long ago passed its mile stone of local repute, and has reached out over the state, the nation and across the seas, until now it has a world wide reputation. People come thousands of miles to see the Goodnight park, with its surrounding enclosure of many acres, surrounded with an eight wire fence and holding within its limits deer, elk, and the famous Goodnight herd of pure blood buffalo, as well as half breeds. When I made my last visit to the Goodnight home the herd contained about 130 head, seventy-nine of which were pure bred buffalo.

When Mr. Goodnight first located in Paloduro canyon, buffalo hunting was still in full sway and there yet roamed over what in the past has been known as the "staked plains," vast herds of these noble animals, but the fall and winter of '77 and '78 closed out forever these great herds, and in the spring of '78 there were only a few scattering bunches left, and Mrs. Mary Goodnight suggested to her husband the advisability of preserving to Texas and the nation a few of the buffalo. Acting upon this wise suggestion, Mr. Goodnight hunted up a bunch and roped two calves, fortunately a male and female, and a few weeks after this Mr. Lee Dyer, a brother of Mrs. Goodnight, captured two more, also a male and female. These were all conveyed to the ranch and each one given a milch cow to suck. They soon became tame and were allowed to run with their adopted mothers on the range with the rest of the milk stock. Here was the beginning of the present Goodnight herd which has not only proven of great commercial value to him, but has preserved to Texas this noble animal, and made the Goodnight ranch famous the world over.

PIONEER DAYS IN THE SOUTHWEST

CHAPTER II. PERSONAL REMINISCENCES. BY EMANUEL DUBBS.

BY request of that typical old pioneer, John A. Hart, and others, I will add a few of my own experiences of the early pioneer days in Southern Kansas, "No man's Land," and the Panhandle of Texas. I have a delicacy in doing this, because of the letter "I" for it may appear to the reader too much like self-exaltation, but I do not know of any way to avoid it, do you? For this is to be my own experience that I am to give you, of course always in connection with others. I am glad that I have an opportunity to do this, because many who participated have already passed away, and soon the little band who blazed the way amid innumerable perils and hardships, will only be a memory.

Their deeds of heroism and self-sacrifice, their unconscious, whole hearted liberality, true and brave hearts beating under rough exteriors, amid wild and rough scenes. They gave their all that a grand fertile country might be reclaimed from savagery and converted into happy homes, waving fields of grain, schools and churches, houses, railroads, towns and cities. Oh, grand and heroic band! I am glad that in an humble way I can be numbered with you. Comrades, I greet the living and honor the dead. My Dear Mr. Hart:

I do not think you or the readers of your book will be interested in my nativity or my four years civil war experience, suffice it to say, that I was born in Stark county, Ohio, on a farm. I knew President McKinley when he was a clerk in a store in our county seat (Canton). When I was eighteen years of age I entered the union army as a "high private in the rear rank," in the First Ohio Volunteer infantry, Company I. I served four years, until peace was declared, when I returned home, somewhat the worse for me, having had all the war experiences I wanted, having been wounded at Stone river and again at Kasaca, Ga.

My welcome home by my gray haired old father and my loving Christian mother, was that of the prodigal's return. My best girl, I think, was also glad to see me, and to reward her I married her, for which kindness on my part, she says she has been sorry ever since. (Oh, the ingratitude of the human heart.) However that may be, she

1

has been my "help mate" indeed. For all these years and through every happiness, through all trials and adversities she has been next to my God, my strength and my helper.

In the spring of 1871 we, that is wife and I, concluded to follow Horace Greeley's advice and move west, for we had lost our little fortune in a fire. We made the trip with a mover's outfit, a covered wagon, a pair of mules, small tent, camping outfit, etc., and one evening we pulled into the then outside frontier town of Newton, Kansas, the end of the track of the Santa Fe and Atchison railroad. We did not intend to stop only long enough to purchase supplies, as we had taken a contract of grading work on the line between the Little and Big Arkansas rivers near the town of Hutchinson, Kansas, which was not then in existence. When we entered the town of Newton we found her citizens wildly excited; and on making inquiry as to the cause we were informed that on the night previous a little trouble had arisen in a saloon and gambling hall called the "Gold Booms" between the cowboys and some railroad men and as a result of said unpleasantness there still lay on the floor seventeen dead men, the wounded having been removed. My wife did not want to stay long enough to buy supplies, but I pointed out to her that the dead men could not hurt her and that the wounded, I supposed, had enough of it and were not likely to hurt us, either.

Many interesting things happened while doing contract work on the grade, which in a work of this kind I think do not directly apply to pioneering. I will pass over the time to the spring of '72. Sometime in the month of June, myself and the head contractor, Mr. Wiley, and another gentleman, whose name I have forgotten, traveled up the line from old Fort Larned over what was then known as the Dry Ridge trail. Coming out in sight of the beautiful Arkansas valley about two miles above where old Fort Dodge then stood, a panorama opened up to our view that was most beautiful. The island dotted Arkansas river winding through this green valley, covered with luxuriant native grass, a stream at this time of year on an average of a half mile wide filled with water from bank to bank, with beautiful groves of cottonwood on the islands and along its banks. Having traveled all day without seeing a tree, this was a refreshing sight, indeed, and what made it still more interesting to me was the numerous herds of buffalo feeding contentedly on the nutritious grass, others lying down all unsuspicious of danger, and if the truth must be told there was no very great danger to them, but they did not know it, neither did I, but I thought there was, to one of them, at least. I was riding a mule and was armed with an old cap and ball pistol. I requested Mr. Wiley, who was riding in a carriage, to hold up a little and I would soon get buffalo meat for supper. He consented to wait and I got out my "old Cap and Ball" and put the spur to my festive mule and sailed down on the herd. I singled out a fat cow and as my mule forged up by the side of her, I pulled the trigger and "presto!" behold the change. I think every one of the charges in that old six-shooter exploded at once. My mule made a right "chasse" and I took a header to the left. When I recovered consciousness, Mr. Wiley was trying to raise me up to a sitting position. I lost my buffalo, I lost my religion, I lost my sixshooter, and I thought I had lost my mule, but on looking around I found her grazing contentedly about two hundred yards away, as innocently— well, as innocent as a mule.

We made camp that night five miles up the valley from Fort Dodge, and next day when Mr. Wiley's outfit arrived, we put up a supply house, which was to supply the men, who were soon to follow, with provisions, feed or grain for teams, etc. This house was put up in sections, every thing having been prepared before hand. This was the first house ever put up in what soon afterward became known as Dodge City, Kansas. This most noted of all southwestern towns, noted, for scenes and conditions, and characters that now no longer exist and reads like an "Arabian Nights" romance. I made this town my headquarters for two years in my hunting expeditions south into "No Man's Land" and later still further south into the Panhandle of Texas, about which I will speak later.

I believe at this place it would not be amiss to tell the reader something of the character and conditions that prevailed then—scenes in which I participated and others to which I was an eye witness. Looking back now, after a period of thirty-five years and remembering my early training under the zealous care of a strict Christian father and a mother's unselfish and devoted love, besides the never failing and tender solicitude, and self-sacrificing spirit of my wife, who through every change, and trial, and danger, has always been my helpmate, my companion, my counselor and friend. Notwithstanding all these safeguards, my new environments exercised a powerful influence over me, and it did not take long until I was no longer looked upon as a tenderfoot.

The name of Dodge City, Kansas, was known far and wide, and her reputation was not enviable. The town grew almost in a night into a tented little city, every man was a law unto himself. In a few days "Boot Hill" grave yard was started. At the approach of night the dance hall, saloons and gambling halls were a blaze of light and activity. The sharp report of the six-shooter became a nightly occurrence, and in the morning the usual question was: how many were killed last night?

I will here relate just one event out of the many in the early history of this town which will give the reader a very good conception of the general cussedness that prevailed. Among the many lawless characters, who in the very beginning drifted into town was Billy Brooks. In a few weeks he had established a reputation as the killer. He was wonderfully quick with a gun. He carried two, one on each hip (as did nearly every one else), and in the flash of an eye he could draw one in each hand and fire. In less than a month he had either killed or wounded fifteen men. I did not witness any of the shooting, as I was then and for six years afterward engaged in buffalo hunting, (my camp at this time being on Bluff creek, twenty-five miles from Dodge City, south of the Arkansas,) saving the horns, humps, hides, etc. Late one evening I pulled into town with two loads of buffalo meat (this was in the winter of 1872-73). After putting up our teams and eating supper I went to town into one of the stores to make the arrangements to dispose of my meat and buy supplies so we could get off early next morning. While talking with the merchant, we heard a number of shots fired, and a woman screaming for help in a gun and ammunition store adjoining. I ran out and started for the door of the house where the shooting was going on, and just before I came to the door, Crash! Bang! There came half a dozen shots through the door from the inside. I concluded that was not a very healthy place and went back to the store where I had been trading and asked the proprietor what the trouble was. I found him badly frightened and closing up. Not being able to obtain the information I wanted, I went out again, but the street was deserted. After a little I saw four men come out of the door and start down across the railroad to the south side where there was a dance hall, that at this time of night was in full blast of activity, as this was the nightly resort of such characters as Billy Brooks and lewd women. These men were going in that direction. Having by this time become intensely interested in the conduct of these four men, knowing that their purpose to be evil, meaning danger to some person or persons, I followed them at a short distance to find out what their purpose was, and prevent mischief if possible. Just then I was overtaken by an old comrade and buffalo hunter by the name of Fred Singer, who informed me that they were after Billy Brooks. Just before they entered the door of the dance hall we shouted a warning to Billy Brooks. They fired several shots back at us, which we returned. Then from the door of the dance hall came such a fusillade of shots from the revolvers in the hands of Brooks, who stood in bold relief in the light of the door, that it appeared to me as if a whole company had fired at the same time, and when the smoke cleared away two of the four were dead, and the other two were mortally wounded. Billy Brooks escaped as usual with only a slight wound in the shoulder. One girl in the room was seriously wounded by a stray shot. Afterwards, I learned that the men who made the attack were four brothers from Hayes city. That some time previous, during a dispute over cards in the place and during the shooting Brooks had killed his opponent, Berry by name, and these were his brothers, who came over for the expressed purpose of killing Brooks, and in order to get up steam they had been filling up on bad whiskey, and after shooting out lights and smashing mirrors in the business part of town they at last got what the people thought they deserved; any way there was no investigation, of which fact I was well pleased because of the unintentional mixup I had in it myself.

As stated before, I made Dodge City my headquarters for two years, but as I hunted buffalo for their commercial value, that is after the buffalo moved farther south, I could no longer convey the green meat to market, and as I thought it a great waste to kill them for their hides only, I concluded to make fall and winter camp with dry houses in connection. So early in the spring of 1873 I filed on a hundred and sixty acres of government land at a big spring five miles north of Dodge City on Duck creek and stocked it with dairy cows, to supply the Dodge City market with the products of the dairy, intending in the early fall to start south with an outfit to hunt the buffalo for meat as well as hides. This meat was sugar cured, and in the spring was freighted to the railroad and shipped in carload lots to Kansas City, Missouri, and from there was distributed to the wholesale houses on the different roads running out from there. I have given this explanation of the short break I made in the pursuit of the buffalo to lead up to our experiences with horse thieves or "rustlers," as they were then called. The lawless condition of the country brought together large bands of this kind of gentry.

For more than a week small parties of men with Indian ponies, mules and horses were brought to their rendezvous just below our ranch, almost hourly, until there were gathered together forty-odd men and between four and five hundred head of stock, under the leadership of Dutch Henry and Tom Owens, two of the most desperate characters that ever went unhung in the west. As I knew the leaders well, as well as many of the men—for I had fed them often when they drifted into our camp—I was in hopes they would not interfere with my stock. The law abiding people had not organized against them, for we did not know when we met a man if he belonged to the rustling gang or not, and it was as much as a man's life was worth to interfere with them in any way.

I had, however, taken the precaution to sleep out on the ground several hundred yards away from the house with my big fifty buffalo gun, as well as a brace of big Colts' revolvers by my side. Here I picketed three of my best horses. This was only a precaution, for I did not believe they would steal from me. However, I was mistaken, for on awakening early one morning I found the picket ropes cut and my horses gone, and when daylight appeared I found the "rustlers" and all the stock gone also. I was afoot without any horses or mules and five miles from town. But I footed it into town and by twelve o'clock we had thirty-five determined men, well mounted, a fourhorse supply wagon fitted with provisions, camp outfit, etc. We got together to organize; I was chosen captain, Ed Jones lieutenant, and by one o'clock we took the trail. Before starting, however, we telegraphed to

PIONEER DAYS IN THE SOUTHWEST

Hayes City, Kansas, asking them to head them off, as I had discovered that their trail led in that direction. To make a long story short, we caught up with them in the hills west of Hayes City and after a desperate battle, Dutch Henry was wounded six times, Owens killed and others of the gang killed and wounded. We were assisted by twenty of the Hayes City force. A number of our men were also killed and wounded, but we recovered and captured all the stolen stock and forever broke up the most desperate gang that had ever organized to depredate on peaceable citizens. I want to state here that the leader, Dutch Henry, recovered and escaped and many years afterward when I was county judge of Wheeler county, Texas, he came to my house and stayed two days—a thoroughly reformed man. What became of him afterward? I do not know.

According to our western expression, I am only able, Mr. Hart, to give you my experience in "rough places." To give you anything like a full account of adventures and interesting events that naturally became part and parcel of the occupation in which I was engaged, would make a volume larger than you contemplate publishing, even condensed as I will make it, it necessarily becomes somewhat voluminous.

I believe, however, that it becomes a duty to myself and the many co-workers who engaged in the occupation of buffalo hunting to make some comments, and correct some errors that prevailed in the minds of those who really know nothing of the condition of things that then existed.

The question has often been asked me: "Why did you slaughter these noble animals in such a wholesale manner?" I will answer our critics. That just as long as the buffalo ranged over so much of the most fertile country in all the southwest, settlements were impossible, because it made all this country the home of depredating Indians, who constantly would steal away from their agency where the government clothed and fed them, to kill and steal everything that came in their way. Even cattlemen, the forerunners of civilization, could not live here, as it was entirely too hazardous. Besides the buffalo, and homes for the people could not exist together, hence to the buffalo hunter more than to any other force is due the opening up for settlement; and thousands of happy, prosperous homes now occupy this country,

7

which even yet is in its infancy, in a comparative sense, with vast possibilities in store for the generation now here and the generations yet to come. I am happy in the thought that God permitted me to live and see, at least the beginning, of the great things in store for this country. The buffalo hunter was the true pioneer that made all these things possible.

A short description of the manner of killing this noble game will be interesting to your readers. The general opinion prevails that the buffalo was hunted more on horseback. That was true of the Indian and a few sporting men who drifted out here from the east, but that mode of hunting was not followed by those who made it a business.

I will here give a description of the manner in which I hunted them. Early in the morning after breakfast I would take my big fifty Sharp's rifle (long shell) using 110 grains of powder, my buffalo horse, two belts of ammunition, about eighty rounds, and start ahead of one wagon and team which generally consisted of one wagon and team and four men. Riding up on the windward side of the herd out of sight of the buffalo, jump off the horse when it was no longer safe to ride any closer, drop the reins over the horses' head, then walk as close as possible without frightening them, say three hundred or four hundred yards, then pick out the buffalo that was in the most favorable position, take aim, being careful to shoot just back of the fore shoulder so the ball would penetrate the lungs. The report of the gun would frighten the herd and they would start off at a tremendous speed. The next shot would not be fired at the herd but in front of them, the whistling of the ball and the dirt knocked up would turn the buffalo back, this would be repeated until I got the herd to milling, and until they got the scent of blood of the first buffalo shot. By this time their attention would be entirely distracted and the hunter could shoot them at his leisure. Another question often asked me was there any danger? Yes to the novice there was considerable danger. The buffalo would not attack a man unless desperately wounded and the hunter was close to them. I know of several men that were killed that way. I had a very narrow escape in the beginning of my experience in hunting, I had killed seventeen out of the bunch, outright, but a cow that I had shot several times moved off four hundred or five hundred yards then laid down, I followed her up holding my gun in my hands ready to fire in an instant. She allowed me to approach her, probably within ten feet, when she jumped up like a flash and came for me with a bellow, her little black, wicked eyes flashing. I did not take time to aim, but fired at random. In an instant she was on to me. I had thought out beforehand just how I would act in such an event. My idea was that I could jump to one side and when the animal passed by in the mad lunge I could shoot to kill. But a great many theories do not stand the test of actual service. Neither did this one. I jumped to one side, but this did not avail me anything, for the buffalo turned so quickly that her sharp horns ripped open my hunting shirt and considerable of the flesh on my right side, knocking my gun out of my hands. I grasped her by the horns hallooing: "Huha! huha!" But she did not huha for a cent. I hung on like "grim death to a nigger," my feet in the air while she revolved on her hind feet. All at once she fell stone dead, throwing me several feet away, also nearly dead with fright. The boys who had been in plain view came rushing up to help me. It is safe to say had the buffalo not fallen dead as she did they would have been too late, and this little incident would never have been written by me. In cutting her open we found that the shot I had fired at random had made a small groove in the lower or small end of her heart and when she fell it was her last struggle.

In the spring of '74 having disposed of my cured meats in the Kansas City and other markets and returning to Dodge City I found A. C. Myers and Chas. Bathe just ready to start south to a place known as Adobe Walls in the Panhandle of Texas, The buffalo had quit the range north of the Canadian river, except in small herds, and the object of these enterprising men was to move nearer the buffalo with a large hunter's supply outfit. They asked me to accompany them, which I proceeded to do, breaking over my rule of summer hunting. I fitted up one team of six yoke of oxen with trail wagons and one four mule team, with regular buffalo riding horses. When we started from Dodge City, we had 100 teams all loaded and about 110 men. My outfit consisted of the teams above mentioned heavily loaded and three men besides myself.

We found the "Adobe Walls" consisted of the remnant of broken down and decaying Adobes, possibly constructed by Mexicans, the true history of which is not known. What scenes may have been

9

enacted there in the unknown past. The sight of these old decaying Adobes, brought to mind visions of heroism, of suffering. I wished these old walls could speak and give the true history of the necessity of their being built, who built them? and why? But they were silent mouldering pieces of earth, no one is ever to know anything about them.

We remained here five days, helping to start the buildings that were soon to be the scene of a terrible conflict. The buildings were constructed out of logs twelve feet high from the bottom of the trench which was three feet deep, leaving the pickets nine feet out of the ground, the roof made out of straight poles, the ends resting upon a center log, called a ridge lig. Then the top closely covered with fine brush, and earth put on top of that. A large corral was built adjoining also made out of pickets. On top of the supply house or store was built a look out. Another building was used for a saloon, of which I will speak later.

After unloading most of the supplies, we only kept enough to last four men about a month, with grain for the mules and horses. In fact about enough for one month's hunt. On the morning of the 6th day after reaching Adobe Walls we bid good bye to our friends and started south, crossing the Canadian at the mouth of White Deer, traveling up that stream to the head of it, to where it opened out on the plains near where the town of Pampa is now located. Remember this was entire new country to us. Very few white men had ever been over any portion of it; there were no trails or roads, only trails made by the buffalo, which led in every direction, often worn a foot deep by the countless thousands of hoofs that passed over them. For centuries this had been the home of the wild plains Indian, the buffalo, the deer and the antelope. In the timbered creeks were thousands of turkeys and quails, and a little further east in the sand hills were innumerable prairie chickens. It was indeed the hunter's paradise. In the four years I hunted and scouted in the Panhandle country her rich and fertile valleys, her limitless prairies rich in soil beyond comparison appealed to me with her great resources, now given over entirely to the Indians, the war dance and tribe wars, but all of them at any and all times ready to murder, burn and mutilate their natural enemy, the whites. I sometimes read sickening rhapsodies about the noble red man. But never from the pen of a bona fide pioneer who has witnessed their unspeakable cruelties. They are a lazy, dirty, lousey, deceitful race. True manhood is unknown, and they hold their women in abject slavery. The most contemptible name they can give a man is to call him "squaw." They are the only race of people under the sun that can not be reclaimed by Christianity, and civilizing influences. The old adage "The only good Indian is the dead Indian" is not very poetical, but it is true.

I must ask pardon of the reader for this digression. We moved southeast after striking the plains, taking our direction from an old imperfect map, expecting to strike the head of the North Fork of the Red River, but we were too far west. The natural depressions on the prarie were now only dry lakes without a demo of water, with the deceptive mirage. All day—and just as the sun was disappearing and night coming on, we came to a very slight depression which indicated to the practical eye the beginning of a creek or water course, but how far off, of course, none of us knew. I ordered the teams taken out to rest, giving strict orders that a close watch must be kept I had one of my men follow the depression in search of water for which we and the teams were positively suffering. After riding probably five or six miles we found a depression or hollow in a table like rock, containing possibly two barrels of water, which the sun had not yet dried up. To make a long story short, we moved the teams down to it that night and next day moved down and crossed this stream, for it was a stream, which we afterward learned to be Salt Fork of Red River, and towards the second night we came in sight of a vast herd of buffalo at a place now known as Lelia Lake, a beautiful little body of water, covered with many kinds of water fowls, as well as many varieties of fish in the lake. It made an ideal summer camping place, and we were not long in preparing it. By the following morning we were ready for business, that is, killing buffalo for their hides which had a market value from $2.00 to $2.50 each, according to quality. We spent three weeks at this place, until we had over a thousand hides. We loaded our teams heavily with these hides that were thoroughly dried. After securing the remainder by tricking them in piles, we started back in the direction of the Adobe Walls, intending to renew our supplies and ammunition and procure freight teams to haul in the remainder that we had to leave behind, while we continued to hunt. As I stated before there were no roads. We pursued a general direction, carefully picking out the best road, or rather ground, we could find. Late that night we made camp in a grove of cottonwood trees, where a beautiful little stream emptied into a larger one. I want to be particular in describing this place, because here was enacted one of the most terrible experiences of my whole life. On the south of our camp just at the mouth of the smaller stream was a high bluff, higher than any just immediately around it, though the whole country was more or less broken into hills and ravines and small valleys. I will state here for the benefit of the reader, that the creek just mentioned was the mouth of Barton, where it emptied into Sadler, situated about ten miles north of Clarendon, Tex. The public road running north to Allenreed passes through it. Hundreds of people have traveled over it since, never dreaming of the tragedy enacted there.

We had heard rumors of an Indian outbreak, but we had heard them so long and nothing of very serious importance had happened, that we became more or less careless. We were always reasonably well prepared to meet any attack, if made. We as usual turned all our stock loose with the exception of one of the best of my riding horses which I picketed close to camp not from the fear of danger however, but for the purpose of rounding up the stock in the morning, so as to get as early a start as possible. Next morning as soon as it was light enough to see, I told my men to prepare breakfast while I proceeded to the top of the bluff or hill, that I mentioned before, in order that I might get a more extended view of the surrounding country, for I was unable to see the stock we had turned loose the previous evening. But I had no better success from my new view point. I came to the conclusion that they had strayed off into some of the nearby ravines out of sight and was not alarmed in the least. After breakfast I saddled my horse, telling the men to have everything ready when I returned, which I expected would be in a short time, I never saw my men alive again. I hunted out ravines and valleys hour after hour, but not a sign of the stock could I find. I found several places on soft ground, horse and mule tracks only to lose them again. Along about noon I did discover a plain trail about three miles from camp in the sand on Salt Fork of Red River, and to add to my uneasiness, I discovered that they were being driven. I at once came to the conclusion that a raiding band of Indians had driven them off. The trail was plain and I followed it ten or twelve miles when I knew that it was useless. To add to my uneasiness, I feared that the rumor of an Indian outbreak was true, for we had heard such rumors all spring. I concluded to return to camp, warn my men, and make the best arrangements we could to save what we had left. It would be hard to describe to you just how I felt as I returned toward the camp I had left that morning. Visions of Indian cruelties came to my mind as I rapidly covered the ground, riding most of the time in a long gallop. My gun in my hands ready for instant action, avoiding as much as possible, all places that looked favorable for an Indian ambush.

The closer I approached our camp the more careful I became, and just as the sun was disappearing in the west. I came to the foot of the hill or bluff that I had used for a lookout that morning, and which I described before. I had not as yet seen a sign of Indians only the pony tracks of those that drove off my stock. I left my horse at the foot of the hill on the south side, throwing the reins over his head. I proceeded afoot to the top of the hill in order to get a good view of the camp, which as I stated before was about 300 yards away. As I reached close to the summit, I went down on my hands and knees and crawled up high enough to carefully look over without exposing myself.

A sight met my eyes that even now after all these years makes me shudder. At first, I thought the camp was deserted, as everything was so quiet. My wagons were in the same place I had left them that morning. On a closer look I discovered under the last rays of the setting sun, that the tongue of the lead ox wagon was propped up with an ox yoke and across the tongue in a spread-eagle fashion was the naked body of one of my men. Even from where I lay, I could see that he had been tortured to death, for there was a burned place on his breast where the cruel monsters had driven splinters under the flesh or skin, and then set fire to them. I could see nothing of my other two men. I felt reasonably certain that they also were killed. I now looked for Indians. I could plainly see the tufted heads of Indians, under the low bank of the creek, and I knew they were lying in wait for me and as I looked up the stream on the north side in a kind of a pocket I could see a herd of Indian ponies. The sun had now set, and the

13

thought flashed across my mind, will I live to see it rise again? In a moment I thought out my line of action. I crawled back far enough so that I could stand upright without being discovered by the Indians below. Yet I knew that from somewhere sharp cruel eyes were watching me. I walked leisurely down the slope of the hill, when I reached my horse I quietly mounted him. Being careful that my gun was ready for instant use, I pulled my six-shooter around handy to my hand, for I had determined to die fighting; anything was better than torture. I had very little hopes of ever coming out alive. My plan was to ride around the foot of the hill, cross the stream about three hundred yards below the camp and where I had also noticed a short hollow with timber on the opposite side of the main stream. As I came around the foot of the hill I entered the main stream in plain sight of camp. I still rode along leisurely in the direction of the timbered hollow, in an angle away from camp, thinking I could make the Indians believe that I was simply looking for the lost stock. But I had hardly rode one half across the stream when the Indians guessing my real purpose, giving short fierce war whoops, commenced firing at me. I gave my horse the spur, guiding him with my knees, firing at them with my "big fifty" as fast as possible. I soon entered the hollow, having gained some advantage over them, and so far unharmed and out of sight for a little while of the first Indians who were on foot and when I came out of the flat at the head of the hollow, I was all of 600 yards in the lead of those on horse back. I again adopted my first tactics, guiding my horse with my knees and firing back at the pursuing Indians. Their guns not being long ranged, they did not reach me, while on the other hand I saw my shots take effect as several of their ponies dropped, which caused them to scatter, but still pursuing. Of that ride, the unearthly yells, the reckless exultant desire to kill, that had taken possession of me, and for a short time, at least, all sense of personal harm left me. For years afterwards I would in my dreams re-enact that wild ride. Its soul-stirring vividness is yet as fresh in my memory as when it occurred. I did not seem at the time to think that I could escape, but I determined to sell my life as dearly as I could. As I afterward reasoned it out, my life was only preserved because darkness came upon us, and winding through the hollows and brakes that were in my course (for I have ridden over the country many times since) and the excellent qualities of my horse were all factors in my aid. I was steering a general course for White Deer creek and the Adobe Walls. I think now that I came out sometime after the last Indian yell had died on the night air, somewhere near where the town of Jerico is now located, for I traveled that night over many miles of level plains. Along in the night I again reached a rough country, which I believed to be the White Deer creek, and I rode down its course for what seemed to me many miles. My horse laboring heavily, I myself "all in" as we express it (a western phrase.) But I hoped my horse would last until I could reach Adobe Walls, but vain hope. It was about three o'clock when he commenced to stumble and at last to fall to rise no more. Poor noble beast, how my heart ached for him, I pulled off the saddle, as tired as I was, carried it on my back for I knew I was only a short distance from the Canadian and on the other side was safety, the Adobe Walls and friends.

That night walk seems to me like a dream now, every hundred yards appeared to me like a mile but at four o'clock in the morning I staggered up to the door of the first building I came to. It was the building which was erected for a saloon, of which I spoke before. I will here state that it was constructed to resist an attack, having port holes to shoot out of, and situated about one hundred yards from the large store building and no obstructions between the two. After pounding on the door and calling out to open I at last succeeded in awakening those within, and great was their surprise and fervent their sympathy when they heard the news I had to tell them. There were nine men sleeping in this building at the time I arrived there. The most noted of them all was Billy Dixon as we familiarly called him, a great hunter, an excellent shot, a faithful friend, true and brave. He is living yet and if he reads these lines I want him to know, though I have not met him from that time to this, I have not forgotten him; I loved him then for his many excellent qualities of heart and hand, and time as it passes only endears him more to me as old memories crowd around me. But not to consume time and space in recalling his many great qualities and of others of that little band, reader, and you who have homes now in this wonderful fertile and productive country, and the thousands more who will soon come and reap its rich advantages you should erect monuments over the graves of such men as Billy Dixon when they are dead, and better still, you should honor and esteem them while living. If you will pardon this digression I will proceed with my narrative of realities. While "the boys" were preparing me something to eat I was telling them of my day and night experience. I believe now, we believed then, had it not been for my arrival that night and awakening the men so early, the history of the Adobe Wall fight would have had a different ending.

While I was eating, as only an almost starved man can eat, the morning dawn was lighting up, and breaking up the shadows of the most eventful night of all my life. But, dear reader, it was not ended yet, for with the approaching dawn also came the most blood curdling yells, as if a legion of the evil one, had broken loose, accompanied by a fusillade of shots, and as we hastily closed and barred the door and buckled on our ammunition belts, and then loaded shells into our buffalo guns; we saw as we looked through the port holes, the store or supply house surrounded by about four hundred Cheyenne Indians yelling like so many demons. We at once opened fire, and mind you it was no random shooting for cooler and braver men never drew sight over a gun barrel. The fight did not last long, it seemed to be only a few minutes, possibly it lasted thirty minutes. Nearly every shot we fired found its mark. To me it was a positive delight and every shot I fired I thought "there is one for the poor boys that so cruelly lost their lives the day before." The history of the fight has been written before. I will only state here that an Indian will never let one of his number fall into the hands of his enemies, their religion forbids it. But we made it so hot for them that they soon retired some distance, but they left seventeen dead warriors lying on the ground around the stockade of the store. Three of the men there were killed out of seven; among them were the two Shidler brothers, cousins of my wife, who had just arrived the day before with a large supply of freight from the railroad, Dodge City, Kansas. In giving these details the reader may notice that I have not given dates. The reason for that omission is: I am writing from memory entirely and I do not now recall the exact dates. I believe, however, in the main this narrative is substantially correct.

When I commenced writing these reminiscences, I only agreed with Mr. Hart to furnish about five thousand words, and as I have written more than that now, I cannot go farther into this history. My scouting experiences began under Gen. Miles' expedition against the Indians during the summer of 1874. Nor can I give further details concerning the siege by the Indians of the Adobe Walls and the many other incidents of interest connected with it. Nor the many interesting events that occurred during my continued buffalo hunting which lasted until the spring of 1878 when wife and I moved to Wheeler county, among the first settlers. Neither will I have to write about the adventures and the progress of the early settlement of this Panhandle country.

Wheeler county was the first to organize in the spring of 1879 and the citizens elected me their first county judge. We had then attached for judicial purposes, twenty-six counties. Suffice it to say to Mr. Hart and you kind readers, that if you have the patience to wade through what I have already written I will be surprised. I desire to add in conclusion that this great country, the Panhandle of Texas, which was virtually reclaimed from savagery and opened up to settlement by buffalo hunters and pioneers will teem with thousands of happy homes, and that once in a while you will give a kind thought to the memory of those who braved so many dangers, blazing the way for future progress and greatness.

PIONEER DAYS IN THE SOUTHWEST

PIONEER DAYS IN THE SOUTHWEST

CHAPTER III. PERSONAL REMINISCENCES—Part II. By EMANUEL DUBBS.

AFTER sending in my manuscript to Mr. Hart, he requested me to write for him some more of my frontier experience which he believed to be valuable to the coming generations and interesting to both old and young of the present. In my boyhood days, I remember how eagerly I listened to the experiences of early settlers, as of evenings we gathered around the wide open fireplaces of our homes, and they vividly portrayed encounters with Indians, bears, panthers and wild game of the forests; how they subdued those forests, their hardships and trials. I did not think then, enjoying the peace and comforts of a home, that my father and mother had hewn out of the wilderness, that it would ever be my lot to encounter even more perilous experiences, and greater hardships than had these old pioneers, whose stories were so interesting to me then. Yet such was the case.

In closing up my last chapter, the last of which related to the Adobe Walls fight, I failed to give any details of the siege that followed and many incidents connected with it. I will state now that it was probably fifteen days before I could get a party of men to go back with me to where the Indians had killed my companions on the mouth of Barton Creek. And when we arrived there we only found bones, as the prairies wolves had torn and scattered them in every direction. These, however, we gathered carefully, and buried them on top of the hill which I had used as a lookout as described in my last chapter. The wagons and buffalo hides were all burned and destroyed so that as we did not save anything. We returned to Adobe Walls, and from there under a strong escort, we abandoned them and returned to Dodge City. I think some incidents that occurred here during the siege will be interesting to the readers.

There was a hill or small mountain at a distance of 800 yards from the lookout on the store building, and about half way up its steep acclivity was a table rock and just above it against the side of the cliff a white chalky substance. Before the siege and attack, the hunters for amusement used to shoot at it for a mark, some one having stepped the distance which as before stated was just 800 yards. Several days after the engagement an Indian climbed out upon the rock, believing no doubt that he was out of range of our guns, and in order to show his utter contempt he turned his back toward us making indecent motions. Billy Dixon picked up his "big fifty Sharp's," took careful aim, and fired and Mr. "John Big Indian" doubled up and rolled end over end down the mountain. They never tried the experiment any more.

Some ten days after the battle, Mr. Tobe Robinson and a Dutchman, whose name I have forgotten, and probably never knew, we called him "Dutchy," getting tired of being so closely confined, decided that they would ride out on a little scout, for we had not seen a sign of an Indian all day, and thought probably they had raised the siege. Down along the banks of the Canadian were a great many black currants, which were ripe, so not seeing any Indians they filled up on the fruit, and leisurely returned to the "Walls." They had come back about half the distance when about 100 of the Indians rushed down over a sand hill to cut them off, and no doubt expected to capture them and torture them to death afterwards. Tobe Robinson at once opened upon them, while Dutchy equally as well armed, forgetting about his guns, pumped all the wind out of his horse, which soon fell behind Robinson, who guided his horse with his knees and shooting rapidly made his escape in safety without a scratch, while poor Dutchy was speared off his horse and killed. A party of hunters from the "Walls" trying to make a diversion on foot and firing at long range I think helped Robinson to make his escape. We all believed that if Dutchy had shown equal nerve, he too might have saved his life. Nerve and coolness in time of extreme danger always gain the respect of others, and we felt proud of Tobe Robinson. Some ten days after this, we were all escorted back to Dodge City by three companies of U. S. cavalry from Ft. Dodge.

When we returned to Dodge we found all the hunters over the whole range congregated there. The whole valley for several miles was dotted with hunter's camps, of all kinds, and descriptions. Some with tents and others with wagon sheets for protection. Had you, dear reader, seen the motley crowd of men, unshaved and some of them unwashed, you would no doubt have judged them to be desperadoes and outlaws, and you my gentle lady readers would hardly have chosen one of them as your escort on a dark night. But speaking from positive knowledge, there never has been from the days of the crusaders and chivalry, to the present moment, there never has been a better, braver, nobler set of men lived. They were God's true noblemen, loyal, brave, honest and true. A woman's honor as sacredly preserved as if she had been in paradise. Such a thing as petty theft was entirely unknown. You could leave any kind of property exposed and unprotected, and it was as safe as if locked up in a New York city vault. Every man wore a belt, was armed with a knife and two Colts six-shooters. Quick to resent an insult, quarreling almost unknown, for such a thing as a fist fight rarely if ever occurred. If trouble did come up they used the gun and not the fist, this made a man careful how he treated his fellow men. I have given you this description, because these men were true pioneers that subdued the Indians, and opened for peaceful settlement the most fertile and productive country west of the Mississippi river.

The government was preparing an expedition under the command of General Miles against the Indians, but they were slow with their preparations and in mobilizing the troops, and being cooped up doing nothing, did not suit me in the least. About this time Mr. Peters, an old buffalo hunting friend, who owned a good outfit, and knowing that I had lost mine, proposed that I join him as an equal partner. That we hire two more men and go after the buffalo. I accepted the proposition and we found two men who were willing to take chances at $75.00 per month and found. Now this may seem extremely foolhardy to the reader, looking at it from this distance and from my present viewpoint, it does appear so now. But we knew that the Indians were desperately afraid of buffalo hunters, and that there was really very little danger, unless they found us careless and unwatchful. Anyway we concluded to take chances. So we prepared our supplies, ammunition and all necessary for about a three weeks' hunt.

PIONEER DAYS IN THE SOUTHWEST

We crossed the Arkansas river at Dodge City, twenty-seven miles to Crooked creek, then thirty-five miles to the Cimarron river, then twenty miles to the Beaver, across that about fifteen miles to a water hole on the plain between Beaver and Wolf creek, what was then known as "No Man's Land." Here we found the buffalo by the thousands and we had not seen an Indian on the whole trip. Yet we knew that this was no sign they had not seen us. In fact we felt reasonably certain that they were watching our every move and only waiting an opportunity to wipe us out. On this level flat was the safest place to make a camp that we could select, and as we stood guard every night, turn and turn about, we were prepared at all times to get as many of them as they could get of us. We worked hard and in five days, we had over 200 hides, and in a few more days we had them dry enough to haul. By putting four horses to a wagon we were able to haul them at one trip. We got our loads ready to move by noon one day, intending to make Beaver creek by night. Now in order to have the reader understand the trouble we got into, it becomes necessary to make some explanation in regard to loading hides on a wagon, in order to transport them long distances, and how we had to prepare them. First they were stretched out on the ground, hair side down when green, then pegged out with little stakes previously prepared. Then before they became flint dry, we would take them up, fold them with the hair side in, and load them in the wagon with the back of the fold outside and the legs laping on the inside, then tie them down firmly with ropes. By being careful a load could be hauled a long distance without slipping. But in this instance this was not done in that way, because we were hurried so, and we had to load them flat, as they were too dry to fold, and this led to our undoing. When hides were loaded without folding, it was almost impossible to haul them without slipping, and it made it necessary to tie them down with what we called a boom pole, or long springy pole. But we had no poles or timber to make them, so we tied with ropes the best we could until we could reach timber, near Beaver creek. Our loads kept slipping until we came within one and one-half miles from a dry creek that led down to Beaver, which had beautiful groves of cottonwood growing on it. The creek was afterwards named Peters creek. We were compelled to make a halt as one of our loads toppled over, so that it became necessary to load it all over. Now we knew it was dangerous to separate, though we had not seen any signs of Indians on our whole trip. Mr. Peters proposed to me that he would take a horse, singletree and chain and go to the timber and get the poles, while I with the hands, reload the hides. We had quite a dispute about it, as to which one should go. It was a bad job to reload the hides, and Mr. Peters said because of his sore hand (he had it cut a few days before) he could get the poles easier than reload the hides. I at last consented for him to go, and this decision no doubt saved my life. Great events often are the result of what seem to us as very trivial circumstances.

When Mr. Peters left us that evening jolly and laughing, little did we think that the separation was forever, so far as this life is concerned. But such was the case, none of us ever saw him alive again.

The wind was blowing almost a gale, which made it very difficult to handle the hides, but at last we were ready for the poles, and we commenced to think it was time for Peters to return. It was about 200 yards to the brow of the rise from where we had halted, and from there, was in plain view of the cottonwood grove. I went up to the top of the rise to see if I could see anything of Peters, but I could not see him, but in the direction in which I was looking I saw about five hundred yards away what I took to be a buffalo carcass skinned, as it looked white, but at the time I thought nothing strange about it. I went back to camp, caught up a horse and saddled him, and told the men to have everything ready and when Peters came back, to pull for the crossing on Beaver creek. That I was going to an old camp ground where we had stayed over night in coming over, and where I had left a beautiful Mexican made quirt or riding whip, which I valued highly. That I would join them again at the crossing. I rode along leisurely and had gone perhaps five or six hundred yards, when I saw something white hanging to a prickly pear bush. I jumped off my horse to see what it was, and when I picked it up I discovered it was a letter addressed to Peters. I stated before that the wind was blowing hard, and from the direction Peters had gone. The moment I looked at the letter I knew Peters was killed, and the reason I knew it was that Mr. Peters had confided to me a sweet secret. He was engaged to a young lady in Wichita, Kansas, and this was to be his last trip, as he intended to marry her on his return. He carried these letters in his inside vest pocket in a silk oil cloth case. For a moment I was so shocked, and horror-stricken, I was unable to move. But then the consciousness of our danger rushed over me, and I jumped on my horse and started back to camp, taking the stock with me that were grazing a little way from camp. As I drove them up I ordered the men to catch them, and by moving the wagons side by side tying the horses between them then stretching ropes across both ends and piling hides against them, made a protection that would resist any ordinary attack. When this was completed I told the men to guard it at all hazards. As I intended to find out what had become of Mr. Peters. They begged me not to leave them, urging that the Indians would get me also, and then they too would lose their lives. I admitted that there was much truth in what they urged, yet I simply considered it my duty to find out what had become of him. I did not have the heart to go back, (that is providing we ever did get back) and say I did not know what had happened him, and that I did not have courage to at least try to find him. I said courage. Yes, it did take courage to ride by myself through the brakes to the timber, passing a dozen or more places where the Indians could have lain in ambush, not knowing their force, only certain of one thing. Peters my noble-hearted companion is killed. "Peters is killed!" this seemed to me to sum up, and swallow up everything else. I do not claim that I had courage above the ordinary, yet I had in the past discovered one trait in myself, that the greater and more imminent the danger the more determined I became to face it, and the steadier became my nerves. So it was in this case.

I first took the precaution to tighten the cinches on my saddle, pulled both my six-shooters around handy to my hands, my belt full of ammunition, my gun across the cantle of my saddle, my right hand grasping it so I could shoot instantly. Avoiding all I possibly could, all dangerous places, where they could lie in ambush. In this way I made my way down to the timber without seeing a single Indian. When I arrived in the timber I soon discovered the two trees he had cut down, and there was a plain trail where he dragged the poles through the sand up the sandy bottom. I could see the trail all of a hundred yards ahead, I followed it at a lope. The birds chirping and whistling in the trees and in the thickets were suspicious sounds to me, and every thicket I passed I expected to feel the bullet or the arrow. But nothing happened. In a little while at the head of the last draw or hollow, where I thought laid the buffalo carcass, I found poor Peters. Also in the hollow a few rods away a dead Indian pony. He, that is Mr. Peters, was laying on his face, his clothes torn off of him, scalped and his ax sticking in his head and otherwise horribly mutilated. While they had evidently taken him by surprise, yet he must have made some kind of a fight because on examination I found the Indian pony shot. While Peters was shot in at least three places. But I did not lose much time in my examination, but hurried back to camp. Here I found my men anxiously awaiting. I proposed that we bury Peters, but the men refused positively. We only had an old broken spade that we used about our camp fire in cooking and the buffalo grass turf was very hard. It really was impossible to dig a hole large enough to bury him in, so I gave up the point. The next question was the best and safest plan to get out of there, and if possible, back to Dodge City. We did not yet know what force our enemy had. But one thing I did know that they never would attack us openly, unless we gave them every advantage, and this of course it became our study to avoid.

I told the men to throw about half the hides from each wagon, in order to lighten the loads. It was good policy to keep as many hides as possible for protection. No bullets that the Indians had at that time could penetrate two thicknesses of a flint dry buffalo hide, for I have seen it tried, and I will give a description at another place.

After having the loads remodeled to suit us, we hitched our horses to the wagons. We now only had three for one wagon, for the Indians got away with the one that Peters had taken with him. It was probably an hour until sun down when we got started. We had some bad rough country to go over until we crossed Sharp's creek on the other side of Beaver. Everyone had his gun and ammunition ready for instant use. I rode ahead from one point to another, keeping a sharp lookout so that we could not be taken by surprise. We traveled very slowly, and carefully. It was probably one o'clock in the night when we gained the flats on the other side of Sharp's creek. Here we made camp for the rest of the night, all of us badly worn and tired. Yet it was necessary for one to stand guard. I took the first turn, and the last, for I was afraid to take chances, so I watched till nearly daylight, when I awakened both the men, and told them to prepare a light breakfast while I laid down with my clothes on. I did not remove my belt of ammunition and six-shooters; that however did not bother me for I went to sleep at once, and the sun was an hour high, when the men awakened me, with the cry, Indians! Indians! Sure enough not over half a mile away from us riding along single file were twenty-two Indians going in the same direction that we were headed for, the sand hills on the south side of the Cimarron river. I did not have the least doubt that they were the same party that killed Mr. Peters, and had been watching us, for a favorable opportunity for an attack. I fired several shots at them, which caused them to hasten, but I did not get any of them. This was the last we saw of them, but not the last we heard. Right on the Crooked creek flats they killed six government surveyors. Then they run into Fred Singer's hunting outfit on the road from Crooked creek and Dodge City. They circled the hunters shooting at them from under the horses necks, but this fight was soon over. When I came along next day, there laid in that circle twelve Indian ponies. I took one of their saddles, made out of the fork of a tree, which my boys used afterward as a novelty. I do not know how many Indians were killed in this fight. Singer had three men with him, they threw themselves flat on the ground and made every shot count. We arrived at Dodge City without, any further adventures. I notified Mr. Peters' people and bought from them a team and wagon.

Probably ten days after my return to Dodge City, Tom Nixon, a famous hunter and scout, came to me with a proposition that we fit up a large outfit and kill buffalo in spite of Indians or anything else. Finding out that it would be yet sometime before the troops, who were concentrating at Ft. Dodge under General Miles, would be ready to start on their campaign against the Indians, I accepted the proposition, and at once helped to organize an outfit. It did not take long to find the right kind of men among so many adventurous spirits as were assembled around Dodge City. When we started we had twenty-eight men and fourteen wagons and teams, with our riding horses. When we made our first camp on the south side of the Arkansas river, Tom Nixon was elected captain and myself second in command. We pulled out on the same road that my companion and myself had so lately come in on. But I did not take either one of them with me, as they were too much afraid of Indians to suit me. I do not think they would have gone if I had asked them, for I think they had all the hunting they wanted. This trip we did not find the buffalo in large numbers, until we went beyond where Peters and I had our camp. But not quite in the same direction. We finally found great herds on the flats between Wolf creek and the Canadian river. We made our camp at an inland lake made up of rain water, mud and the urine from the thousands of buffalo that came there to drink. It was impossible to drink it without first boiling, then make it into coffee. About the third day we found that our outfit was too large to make hunting profitable, So it was decided to divide the outfit into two equal parts. Ed Jones and myself were elected first and second in command, and we proceeded at once to make a new camp. It was probably ten o'clock in the morning when all arrangements were completed and we moved out intending to make our camp at a place known as Big Springs on the head of Wolf creek. We had gone perhaps six miles, when we came in sight of Wet Weather lake, and grazing around it, was a large herd of wild horses or mustangs. There was quite a herd, probably 300 head or more in this bunch. There were five or six herds, roamed over this part of the plains country. There were some beautiful horses among them. All dark colors, brown, black and bays. One thing about them that was peculiar, and was always a surprise to me, and that is that no difference how rigorous the winter or at what season of the year I never saw a poor horse among them: that is thin in flesh. No one knows the true history of their origin. Some writers believe they originated back at the time of the conquest of Mexico by Cortez. As they are now I believe entirely extinct, just how that happened, I am unable to say, and have never heard an explanation. It will be instructive to the reader to explain their habits and the manner of catching them and I will here give you a short description.

There were very few men in this part of the plains that ever tried catching and taming them. Each herd had a certain range, and when followed any distance they would travel in a wide circuit, making twenty-five or thirty miles in a circuit.

The first thing to do was to build a large corral, very strong, in a grove of timber on the ground they ranged over, and to the opening or entrance to this corral large wings or chutes were constructed, say two hundred yards wide at the mouth, narrowing down to the entrance of the corral. Now having the corral constructed, the next thing to do was to get the herd in condition to drive them in. It would take three men with extra mounts to do this. The extent of their range or circuit was first ascertained, then the three men would take up their position separately, as near as possible at equal distance from this circle. The first man would start the horses and follow them on an easy lope, until the horses reached that part of the range where the second man was stationed, when he would follow the same way on to the third while the first and second rested. This was kept up day and night, until the herd became tired, and also accustomed to the riders. When the men thought them in proper condition, all three would ride after them on fresh mounts and drive them into the corral. After having them secured in the corral, the next thing was to rope and firmly secure all they wanted out of the bunch. When they were once conquered and found that they would not be hurt they were as easily broken as range horses. I have made this description just as short as possible in order to give the reader a clear idea of early frontier life, that with the wild buffalo have passed out forever.

With this short digression we will go back to our hunting outfit. When we reached the horses, they threw up their heads, and scampered off to another part of the range. It was a beautiful sight. We noticed one stallion, black and his coat shining, his mane and tail almost dragging the ground. He was the leader, and very proud he seemed to be of his charge. We watched them until they disappeared over a swell on the prairie. Then we drove up to the water and prepared our dinner, as it was noon. After our nooning we again started for Wolf Creek Springs. We had just half of the original outfit, seven wagons and teams and fourteen men, Mr. Jones and myself in the lead guiding the outfit, as there were no trails or roads.

When we approached the brakes, or small tributaries of Wolf creek we started down the ridge between two tributaries and did not go far until in looking down and across on the north side of the main stream, we discovered a herd of moving objects, which we first thought were antelope, but on looking closer we discovered that they were of different colors. Yet the distance was so great we could not tell for certain, but I pointed out to Mr. Jones that they were horses, and that they could not be wild horses, they never had any white or calico horses in their herds. So we both came to the conclusion that they were Indian horses, and also that there must be Indians about. We rode back to the lead wagon, a four horse team owned and driven by a man who claimed that he was an old government scout as well as an Indian fighter. This man I think was a little sore, because he had not been chosen as one of the officers. Anyway when we pointed out, to him what we had seen, he pooh-poohed the idea about Indians. Said if you are afraid you had better get some one to run this outfit who is not. Well, we said, all right, if you can stand it we can. I know that was not the proper thing for us to do, we should have consulted with others and investigated more thoroughly, but what we did do was to order him to move on. We proceeded possibly half a mile, when we saw an Indian dash upon a hill close to the main stream, take a look at us, then dash down again. We knew we were in for it. We rode back to the teams, turned them back on a run for we were on a narrow ridge, which gave the Indians all the advantage, as they could screen themselves, while we were exposed, and we wanted to make the level flat before the attack commenced. Jones and I rode along each side of the teams and with our whips urged them on. The position of the teams were now reversed, our brave "scout" and Indian fighter was now behind. Quicker than I can write, bullets were singing around and past us. The Indians came up the hollows on each side, and in a minute one of our brave scout leaders went down and his team all tangled up, and him yelling for help. We had to turn the rest of the teams back and make a corral with the wagons and hides, some returning the fire of the Indians while the rest worked. By the time this was done five horses were killed and one man shot through the hand. All the men were fairly well protected, but every once in a while a bullet would pass through and hit a horse or mule. Quite a brisk firing was kept up on both sides, without a great deal of execution on either, until about three o'clock a party of fifteen Indians tried to cross the little divide from one hollow to the other. I think they thought they were out of reach of our guns, but they soon discovered that they had another thought coming to them. We waited until they were about half way across. We raised our sights to five hundred yards, for we could figure distances very closely and we guessed right; there were only three horses out of the fifteen ever got across and only five Indians. Those who were wounded crept up behind the dead horses, some two or three fired at us, but not very long, for a dead horse was very slight protection against our long shell "big fifties," and "forty-fives." This put a check to the attack for awhile. About sundown the firing ceased on both sides, and we counted up our losses and consulted what to do. We had lost nearly half our stock. But we concluded by using our riding horses we could make out to move all our wagons, by only hitching two to the wagon. There were none of our men hurt except the man who was shot in the hand in the beginning. We decided that as soon as the moon went down, which would be about 9 o'clock, we would quietly hitch up and drive back to the main camp that night. We did not think we were strong enough to make an attack on the Indians whose numbers we estimated to be between forty and fifty, with a large herd of horses. While we were waiting for the moon to go down, we would put corn in our bread pans and hold it in our hands while the horses ate, because we were so crowded we could not feed them the regular way. I had given mine one small feed and thought I would get some more out of the wagon. There had not been any firing for some time, and the moon only lacked about an hour to setting. I got upon the wagon with one foot, and the other resting on the cross board of the wagon bed, while I was dipping corn out of the sack with my hand, a bullet came so close to my hand it almost scorched me, and the instant flash and report of a gun fired out of the shade in the hollow about seventy-five yards east of us. The shock and surprise was so great that I lost my footing and fell head-first into the ammunition box which was standing open by the side of the wagon. The boys fired at the flash of the Indian's gun, and then helped me out, asking where I was shot. I told them I was only frightened, and not hurt. It appeared that the Indian had crawled up the hollow where it was quite dark and getting my head between him and the moon, he took deliberate aim and only missed me by a scratch.

After the moon went down we quietly hitched up and just as quietly pulled out and at day break we made our main camp. We soon had the whole camp up and listening to our account of the day and night's adventures. As soon as we had breakfast, we picked out our best saddle horses and saddles, and fourteen men including myself and Tom Nixon, whom you will remember was chosen captain in the first start at Dodge City. By ten we arrived where the fight occurred, and here a sight met us showing what the Indians would do to us if they could get us in the condition in which they found our dead stock. They had scalped them, then made checker boards of one side by gashing them with their knives, then turned them over and served the other side the same way. This of course did not hurt the dead horses and mules any, but it made us anxious to run across them again. And this was what we were intending to do, but it was easier said than done.

The Indians evidently did not want us to meet them. They were gone, and they did everything they could to confuse us and hide their trail. It was probably an hour before sundown when we came in sight of them about a mile ahead of us on the south prong of Wolf creek. Now commenced a race in good earnest. I was mounted on a splendid dark brown cavalry horse that a party of us found a short distance from Dodge City, who had in some manner got lost. I could easily keep in the lead. We gained on the Indians so fast that they commenced to drop horses and mules, camp equipage, tepee poles, buffalo robes, pots and kettles, but some time after dark we had to stop the pursuit, as they went down into the Canadian brakes. We did not get close enough to them to get a single shot at them. Having ridden hard all day and covered many miles, you can imagine our surprise, when we looked about us to see our main camp fire not more than three miles away. When we returned to camp the men there prepared us a good hot supper. We needed it, for we had not eaten anything all day. While supper was preparing we planned that at day break next morning, we would take the trail again, and those who remained in camp were to send out a party to gather up the Indian horses and mules and other things of value, that we had no time to bother with in our wild ride after the Indians that evening.

Early next morning we mounted and started out again. A young man by the name of Charley O'Brien had volunteered to go with me when we started on the expedition, with the understanding that if we made anything, I was to pay him, and if not, I did not need to give him anything. He begged hard for me to remain in camp, for he wanted to take my place. He urged that, he had no one but himself and that I had a family, but I would not hear to his pleading. I told him that the men did not elect me second in command to shirk my responsibilities, I introduce Charley O'Brien to the reader, for before the days was out we heard from him again.

When we arrived at the point on the Canadian river brakes where the Indians entered them, we were surprised that they would take chances in the dark to go down such a rough, rocky place, for it was hard for us to follow them in daylight. We found as we followed them mile after mile, winding around the very roughest places they could find, many places so steep we had to almost slide to get down. Toward the middle of the day we commenced picking up horses again, but about three o'clock we decided to give up the chase as our stock was run down, besides we were giving the Indians all the advantage, providing they decided to make a fight, as they could choose their own position. We had left a small party behind to bring along the stock we had captured. So we went back to meet them, and when we all got together again, we hunted out a place that we could get over the cap rock and out on the plains. When we arrived on top of the cap rock, we had an unobstructed view for many miles of plains country. About six miles down in a kind of basin or depression we could plainly see our camp, and to the left about two miles, a sight met us that was exciting in the extreme. Six of our men who had started out that morning, on whatever they could mount, some without saddles, were coming towards camp as fast as they could, driving about forty head of Indian stock, and not more than six hundred yards behind them riding like the wind were a dozen Indians, going two steps to our men's one. It looked to us, that it was all over with our boys, and in our anxiety we commenced yelling and riding towards them, forgetting we were a long distance away from them. It looked to us, as if our men were frightened to death, and we expected that in a few minutes the Indians would be among them with their spears and tomahawks. But such was not the case. Charley O'Brien was in that crowd, and he saved his own life and also that of the rest of the men. When he saw that the Indians were certain to overtake them, he halted the men, who jumped off their horses, had one to hold them, and the other five dropped on the ground took aim and fired. Several Indians were killed and three ponies before they could get out of reach. We all got down to camp about the same time and when we found out that it was Charley O'Brien who had the presence of mind to act so bravely we all lionized him.

We found we had captured seventy-nine head of horses, mules and ponies from the Indians. After letting those who had their stock killed by the Indians, pick out enough to make good their losses, we auctioned the balance to the highest bidder, and when we returned to Dodge City the money was divided equally among all hands. As we all thought we had enough hunting and Indian experience we returned at once to Dodge City, and this wound up our hunting until fall.

General Miles' expedition was about ready to proceed against the Indians and Charley O'Brien and myself contracted to go with them as assistant scouts. This expedition of the 1874 Indian war is a matter of history and the individual bravery and conflicts with the Indians, especially that of Billy Dixon and the three soldiers who were with him, is also a matter of history. There was nothing of particular interest happened to me that I care to write about, or would be of very great interest to the reader, and really this expedition, like all other military expeditions against the Indians, resulted in killing some of them, and forcing the rest to go back to their reservations promising to be good, until such time that they could get ready for another outbreak, murder a number of innocent and unprotected women, children and babies, and butcher a few more defenseless settlers and torture and burn a few others.

Then the government with her strong arm subdues and captures him, always careful not to hurt any more than they could possibly help. Again the Indian would promise to be good, and go like a lamb back to his reservation, and the government would make them a great many presents, plenty of good warm blankets and herds of good beef, in fact everything he can eat, and just as soon as this became monotonous, out Mr. Indian would go, and do the same thing over again. This has been the history of the plains Indians, and thousands of people back east, say, "Oh! the poor Indians, the poor Indians!" And for four years after this every season when I shipped buffalo meat to the eastern market, men would come to me and say: "What makes you kill the poor Indian's cattle?" (Calling buffalo cattle). "The poor

Indian!" I can tell the reader in all honesty and with a good conscience, such talk always made me want to fight, using a western phrase "knock their block off." Of all depraved, utterly heartless and deceitful, dirty and treacherous to the last degree, the plains Indian never had his equal. Poor Indian indeed. All the danger and hardships suffered by the scouts and soldiers in that milk and water system had no permanent effect in settling the Indian question and a permanent peace. Not until the bravest of all pioneers, the buffalo hunters, disposed of the buffalo forever, and which prevented the Indians obtaining sustenance, when they cut loose from their reservations, was the Indian question settled, and their depredations stopped. Not only that, but a grand fertile country was opened up for settlement. The first to see and to take advantage of these great opportunities was the cattlemen. I became personally acquainted with nearly all of those grand, noble, big, liberal hearted people. Such as Charles Goodnight, T. S. Bugbee, Henry Cresswell, Henry Fry, Bill Miller and a host of others who were really the true pioneers of this Panhandle country, and among the stock farmers, was F. E. McCracken, who is a contributor for this book, Tom Bailey, myself and many others who demonstrated beyond a doubt, that this was a fine agricultural country.

Going back again to the buffalo hunters and my own experience, I hunted from the close of the Indian war of '74 to the spring of '78, when I moved my family down to Wheeler county near Sweetwater. The United States government established a military post, and here was the beginning of a real permanent settlement. To give anything like a full account of my last four years' buffalo hunting experience, would take a great deal more space than I can devote to these articles. So I will omit them altogether, and give some description of the difficulties we had to contend with in the early settlement of the Panhandle.

On Sweetwater creek one and a half miles from Ft. Elliott in Wheeler county was built a "hunters supply town" called Sweetwater. It was started early in the spring of '75 expressly to get the trade of the hunters who were operating all over this vast country. And this was the only trading point for nearly two hundred miles north, and two hundred and twenty-five miles south. Here were established hunter's supply stores and saloons, gambling and dance halls. It was a typical frontier town for those days, without law, or an officer of the law.

This town was our objective point, and late one evening about the first of June, we, that is wife and I and our three little boys, landed here, and put up our camp on Sweetwater creek, about 300 yards from the town. After supper I thought I would go up town and renew old acquaintances, for I knew everyone in the town. I just entered Main street, in fact the only street, when in front of Decker's saloon, I witnessed a shooting scrape between Decker and a gambler by the name of Collinson. Decker was shot in two places. The first shot struck on the right side, glanced around under the skin and came out at his back bone, the other hit him, as he fell, above the knee and came out at his hip, both severe flesh wounds, but not dangerous. Collinson was not hurt at all. No one was arrested for there was no one to arrest any one as stated before. But this did not bother wife and I, as we had become accustomed to this kind of life. An old buffalo hunter by the name of Joe Craig had squatted on beautiful sub-irrigated bottom land about eight miles down Sweetwater and I bought his improvements which consisted of a small one-room round log house, covered with earth, and forty acres broke. But it is not my intention to inflict upon the reader tales of our domestic life. Enough to say that our farming and dairy was a success from the start. But it will be interesting to tell you about our method of constructing houses. Remember it was nearly two hundred miles to the nearest railroad and everything that we had to eat and wear had to be transferred all that distance on freight teams.

Lumber cost from $225 to $250 per thousand feet. So in building we did not use any lumber at all. Our first house was built out of tough bottom sod, the walls about three feet thick, and when the walls were completed and settled, they were hewed perfectly smooth on both out and inside. The door jams and facing as well as the windows were hewn out of cottonwood logs After the walls were finished, the next thing to be done was to construct the roof. The gable ends were raised higher in the center, then a large strong ridge log was laid across the center, then plates put on top of the wall, then poles placed very close together for the foundation of the roof. On top of this, fine brush and hay and about a foot of stiff clay was placed and firmly packed. A hard dirt floor, and the house was completed. And it made a very comfortable house. The greatest difficulty was from leaking. Shortly after we had completed our first house and moved into it, we had a long continued rainy spell. About the third night it commenced to leak, until there was no longer a dry place to be found, and becoming very sleepy I told my wife I would creep under the table, it seemed dry there. I did so, and was soon sound asleep. After a while the water partially awakened me, it was dripping on my head. I simply moved my head, but soon the water was dripping on my head again. This continued again and again, until I was thoroughly awake, when I looked around, I saw my wife laughing, in fact I wasn't under the table. She had moved it.

The trouble was, when a dirt roof once started to leak, it would keep it up a day or two after the rain ceased. Every one, however, lived in houses with dirt roofs, and we finally made them so that they would resist every kind of rain. We did not raise much on our farm the first season, and we had to plan some other way to make a living. Deer, wild turkeys, ducks, prairie chickens and quail were very plentiful. I not only made a good living, but I made money, by hunting them, and selling to the citizens and officers and men at Ft. Elliot. In the summer and fall of '78 Sweetwater town was moved from its old site to within half a mile from the fort. This was done in order to get the soldiers trade.

In the spring of '79 the state government granted the petition of the settlers of Wheeler county for an organization, and an election was ordered to elect officers. I had no thought in my mind to offer myself for any of the offices. What was my surprise one day, when an old acquaintance and friend, who had settled five miles north on Gageby creek, came to me, with the news that the citizens had a mass meeting the night before to nominate men to fill the different offices, and that I was chosen to fill the position of county judge. I want to state here, that I filled that position for many years afterwards. For several years, that is until Donelly and Oldham counties organized, we had under our jurisdiction twentysix counties for judicial purposes. Among the first lawyers who practiced in the Wheeler county courts, were such men as Governor J. N. Browning, Col. Grigsby, Hon. Temple Houston, (a son of Sam Houston). I am glad to say, and feel greatly honored, that I made lifetime friends of these excellent gentlemen. There is one incident in the life of Gov. J. N. Browning I think should find space here, and has never been made public. As I stated before, that up to the time of organization, and election of officers, every man was a law unto himself. My readers can well understand that under such condition of things, Mobeetie (the Cheyenne Indian name for sweetwater) became the harbor of the most lawless class of gamblers and blacklegs in all the country, and the law was violated day and night. Our county attorney lacked the nerve, and through policy, shut his eyes to the open violators, and failed entirely to file information and prosecute. The judicial officers were powerless, but a change was to come over the dreams of violators of the law. One day there came to my office a young man of about thirty, jolly, full of jokes, and he could tell the best stories; in fact his whole makeup was a happy, halefellow well met. He presented a commission from Governor Roberts, as district attorney for our judicial district. He visited all saloons, gambling resorts and lawless places of every description, and everywhere he went he was the same happy go lucky young man. It was but a short time, however, until he obtained the evidence he wanted, filed information and swore out warrants for their arrest. When the gamblers discovered that this pleasant young man was in earnest and that he determined to have all violators of the law punished, they concluded to get rid of him. So one day he brought me an anonymous letter, with skull and crossbones, warning him if he valued his life, to leave town and not let the grass grow under his feet. But, that day more arrests were made and in a few days afterwards he showed me another anonymous letter, more threatening than the first one. This began to look serious to me. I told him something had to be done or he certainly would get into trouble and possibly killed. We had plenty of good law-abiding citizens, many of them old buffalo hunters. So by the time he received the third letter with skull and crossbones, twenty-five determined men got together, and I am proud to say I was one of the number. We visited every saloon and gambling house in town, and called all gamblers outside. Then we read a letter to them that we prepared, the substance of which was, that if in any way, or manner, our young friend was harmed or unlawfully treated, that we would hang every gambler and blackleg on the highest tree we could find. This settled it in a short time and there was an exodus of gamblers. I will write his name again in capital letters. The Hon. J. N. BROWNING. I will give one more incident out of many that occurred in these early days, and then I will close this article.

The government brought the northern Cheyenne Indians from their northern reservation and placed them on the reservation with the southern Cheyenne. But the Indians did not like that. So they started and left a trail of blood and unspeakable cruelty along their whole route. Many of the settlers moved to town for protection, but I concluded not to do so, and determined to watch so closely that should they come I could beat them off. One day I was plowing, the lower end of the field was within three hundred yards of the house. I had my "big fifty" buffalo gun strapped between the plow handles, with a belt of loaded shells. In making the rounds, and as I was about half way down to the end nearest the house, I saw a man come over the rise about a mile away, coming on a long lope right towards a spring, where my wife was washing. He had a red blanket wrapped around him, without a hat. I thought certain he was an Indian so I hurried my team until I got within 400 yards of where my wife was, got my gun ready, cocked it and took careful aim over my plow handles, ready to press the trigger had he made any sign of an offensive movement. I don't believe a man could be closer to death's door than he was, for had he made the least suspicious motion I would have pulled the trigger. But halting just a moment to exchange a few words he started off again. I hurried to the house and my wife told me he was one of Millet's men, that they had been shooting up the town, and he was making his escape. I helped to capture the rest of the bunch, but as Mr. Connor, at Memphis, Texas, and Mr. Atterbury at Clarendon, two of our best Panhandle citizens were in the crowd, I refrain from writing more for fear of hurting their feelings. People can reform, you know.

Now, dear reader, I have written at some length, and very hurriedly, and entirely from memory, with only one object in view; and that is, to give true facts of men and conditions, that none of us or the generations coming after us, will ever see again. Most of the grand men and heroic women who, amid many privations, dangers and hardships bequeath to you and yours, as fair and fertile a land as there is any where in all our great domain, have already passed over the river. And the rest of the, old guard, will soon give you their last smile, and press your hand for the last time. I bespeak for them all a loving place in your memory.

PIONEER DAYS IN THE SOUTHWEST

CHAPTER IV. F. R. MCCRACKEN. BY EMANUEL DUBBS

I HAVE before me a short manuscript from E. E. McCracken giving most graphically experiences of his boyhood and young manhood days in Montague county, Texas. I obtained it only because of earnest personal solicitation and because I assured him of the value it would be to present and future generations, and that it would add materially to the value of "Pioneer Days." That which I personally regret more than anything else, is the fact that his self-effacement and modesty prevented him from giving details of his own brave struggles, in the stirring times, when a few heroic men and women stood shoulder to shoulder, and heart to heart, amid the most appalling dangers of the merciless red men, facing dangers, privations, and hardships, that is beyond the power of our imagination to picture or language to describe.

My dear reader, I will take a liberty here for which Mr. McCracken may reprimand me, and that is, to tell you of his nobility of character, his simple, unassuming life, his hospitality and generosity, and the many endearing qualities of heart and hand, that has made him so many friends, who honor him as he deserves, as one of the most honored and leading citizens of the Panhandle of Texas where he now resides, and where I have intimately known him for twenty-five years. The incidents he relates of his early days are not personally known to me, for our lines in life did not draw together until he moved to Collingsworth county which was then attached to Wheeler county for judicial purposes, where I then served as county judge. You have noticed that he says nothing of his Panhandle life and the trials and dangers of its early settlements, and where he and his excellent wife reared their family of boys and girls who now adorn his home and Panhandle society.

It is true that Indian raiding and massacres had about ceased in this section when Mr. McCracken first settled here, yet there were many dangers from other sources and hardships to encounter, obstacles to surmount, that could not have been overcome by any people less inured to frontier life than was Mr. McCracken. It is to such men and their indomitable will and perseverance that we now owe our homes, our school houses, our church houses, whose spires pierce the sky and reflect the glorious sunlight, and from whose pulpits are proclaimed the "light of the world," the great message of love, the "glad tidings of great joy which shall be to all people."

I shall give you word for word Mr. McCracken's Montague county experience, just as he handed it to me. But before I do so, I want to tell you something of his later life, and give some incidents that came under my own observation, that will be interesting to the readers of "Pioneer Days." My object in doing this is, that we, and our children, and our children's children, in the years to come, as they look over this Panhandle country and view its productive fields, its beautiful towns and cities, her thousands of happy homes, her noble Christian civilization, her education and refinement, may in a small measure at least, have a conception of the difficulties and obstacles overcome by Mr. McCracken and his co-laborers. Many of whom have already gone to their long home, some in unmarked graves, and the small remnant yet left, with silvered hair and feeble steps, are looking down the great divide toward the sunset, and soon they too will only be a memory. I wish to say here, that Mr. John A. Hart, the original compiler of "Pioneer Days," and who in this completed work requested my assistance, is himself a typical pioneer, and in the chapter contributed by him and many others, whom he solicited to write for this book, both men and women of excellent repute and honorable standing in society where they are known, yet as you read the trials and danger that beset them, there is not a word of self praise. Yet I can assure you, were their terse sentences, dealing in simple unadorned facts without any coloring whatever interlined, there would unfold to your vision world, pictures of such thrilling romance, of lovers, and self-sacrifice, of immolations of self, and of heroic deeds that would thrill your very souls, and cause you to weave around them the most beautiful garlands of flowers, and the most beautiful of the lady readers would be chosen as the queens of beauty to place the wreath upon their honored memory.

It was, I think in 1880, when I first visited the McCracken home in Collingsworth county, and you who have never been accustomed to such things, can hardly conceive or understand the bare primitiveness of that home. Mr. McCracken had met with severe financial reverses in Montague county and he landed in the Panhandle with one team and wagon. Very few household goods, and without money, His house consisted of one good sized dugout room in the side of a hill covered with poles and earth, called a dirt roof. In fact we all had the same kind of roofs on our houses, so he was in fashion. He had squatted on a hundred and sixty acres of Texas homestead land on Elm creek. He owned a small bunch of cows, and an excellent Winchester rifle and shot gun. After having made his dugout as comfortable as possible with the meager facilities at hand, the question of immediate and future subsistence stared him in the face He had, I think, only one child then, Miss Beatrice, a most winsome child, that somehow had the knack of creeping into the affections of the roughest cowboy on the range. The reader will remember that the nearest railroad point in any direction was over two hundred miles distant, Ft. Worth east, Dodge City, Kansas, north, and the only trading point at Mobeetie and Ft. Elliot in Wheeler county, about forty miles distant, and to give you a slight idea of prices: Flour was worth seven dollars per hundred pounds and everything else in the eating line in the same ratio.

Now take a man in Mr. McCracken's position, with everything to buy and no money to buy with, the reader can, at once perceive the difficult problem that he had to solve. But he was, but a very short time solving it. The country was full of game. The buffalo it was true were gone, and gone forever. Along the timbered creeks were vast flocks of wild turkeys, and the hackberry thickets sheltered thousands of quail. While in the shinnery in the sand hills (little oak bushes bearing acorns) were innumerable flocks of prairie chickens, and in almost any direction, were herds of deer, and Mr. McCracken was a famous shot and hunter. He started out in the fall of the year with his span of ponies and camping outfit. When he found a good location, he made his camp, stayed there until he killed a load of deer, quail, turkey and prairie chickens and then pulled for Mobeetie and Ft. Elliott where he found a ready market. Having disposed of his load, he laid in a supply of ammunition, all kinds of supplies for his family, and then did the same thing over again, until the season opened in the spring. By that time he had enough supplies to last all summer, during which time he raised a crop of feed for his stock, and in the fall he would start out hunting again. And in this manner he laid the foundation, that by perseverance and unremitting toil and economy, has accumulated a handsome competence. I wish to state here that this hunting life was not all sport by any means. Away from his family, exposed to the terrible "norther," there were days of harassing anxieties, and nights of terrible exposure. Whatever the hardships and difficulties or dangers of his life, and, whatever his worries, he was always ready to meet and greet those with whom he came in contact with that affable, gentlemanly manner that has won for him friends among all who know him.

PIONEER DAYS IN THE SOUTHWEST
CHAPTER V. MR. F. R. McCRACKEN'S EARLY LIFE EXPERIENCE

I was born in North Carolina, A. D. 1849 and when I was ten years old my father decided to leave the old "tar heel" state and emigrate to Texas, which filled my heart with joy. I, like nearly all boys, had a streak in me of the hero worship, and a spirit of adventure. From fireside to fireside, from state to state, all over our great nation the names of the heroes of the Alamo had fired the imagination, and hearts were filled with awe and admiration at their heroic death. Texas—the very name filled my young mind with unspeakable delight. My father settled in Willie Wally valley in Montague county. At this time this was the extreme frontier of northwest Texas. For the first few years there were no Indian alarms, everything was peace and happiness in this valley. Nothing interfering with the pursuits and pleasures of the few settlers who located here. It seems to me now as I look back to this time, that the people lived up to the principles of the apostolic days, they owned everything in common. If one family killed a bear or hog, all the families within reach had bear or hog meat. When a young couple married, every man, woman and child for miles around came together to cut down the timber and prepare the logs and in one day build them a house, while the women prepared a feast with such things and eatables as the country afforded, and gave the happy pair a genuine old fashioned house warming.

This happy state continued until the breaking out of the civil war, when all the young men were taken into the army leaving only old men, women, boys and children to protect their homes and dear ones that yet remained from the ruthless inroads of their savage foe who up to this time had not molested this peaceful valley.

In the fall of 1862 we had our first Indian scare. Messengers were sent from house to house, giving the alarm that a band of hostile Indians were coming into our country. My father, Dr. Polly and other old men went to meet them and do what they could to protect their homes and dear ones, but to their great relief and joy they discovered that the alarm was false, for the band of hostiles, proved to be a peaceful hunting party of "Kickapoos." But from now on commenced an era of menace and danger to life and property, that destroyed for years to come the happiness of the people of this beautiful valley. The civil war had taken the flower and best manhood into its death dealing grasp, and as the news of battles fought drifted into the once happy homes, there would also come the sad, heartrending news of the first born, the elder brother, who was slain to return no more forever in this world.

The trouble and dangers increased day by day. The Indians became more and more menacing, frequently we had to "fort up" for mutual protection. The people became accustomed to the danger, and somewhat careless. One day a man by the name of Jones came riding up to our house upon a small played out pony. When he started to leave, my father said, "say Jones, ain't you afraid the Indians will get you riding that kind of a pony?" He laughed and replied: "Oh, I guess I can talk them out of it if they run onto me." But they killed him before he returned home and the valley is called "Jones Valley" unto this day. This killing caused the people to use greater precaution. At the close of the civil war the citizens hoped, for a while that their greatest danger from the Indians was past, but they were doomed to disappointment. The United States government established military posts all along the border, with negro troops. The interior department appointed Indian agents to feed and clothe the Indians, and persuade them to be good, but who ever heard of a good Indian unless he was dead? This old border expression, that the only "good Indian was a dead Indian," was verified to the letter. The agents appointed by the government were corrupt and unscrupulous to the last degree. They traded the Indians arms and ammunition for ponies, stolen horses and buffalo robes, and soon there was a reign of death and terror along the whole border. It was at this time that the United States government commenced the system of placing the control of Indian affairs entirely under the interior department and out of the military, and the result was that the Indians burned and murdered men, women and children. All along the southwest border our once peaceful and prosperous valley became the scene of constant Indian raiding and bloodshed.

While one of my neighbors was returning home from the mill with his family in an ox wagon, the Indians overtook him and killed him, taking his wife, with an infant at her breast, and three daughters prisoners. They had only gone a little distance, when they took the sweet little babe from its mother's arms and threw it into the head of a draw and piled a few stones on its body and then rode off forcing its mother along with them, the cries and screams of her darling baby ringing in her ears. Oh, the agony and terrible distress of the poor mother, her husband killed, her babe murdered before her eyes, and herself and three daughters carried into a captivity a thousand times worse than death. Do you, my dear reader, feel surprised that you never find a genuine frontiersman in love with an Indian? And those sickening rhapsodies of certain Indian philanthropists back east, talking about "the noble red men" become nauseating and disgusting to us who have witnessed their relentless cruelty and treachery? The noble "red men" indeed. This family was in captivity for several years. (This was the Box family.) I am only describing a few out of the many instances, and tragedies that occurred right in our neighborhood. Besides the Box family, there was the Keenan family, and the Huff family, these three families were all killed or taken prisoners, close to my father's home, besides many others that I have not mentioned. The Indians made raids, burned, murdered and stole nearly every light of the moon on up to the close of the war, then our troubles began in earnest.

It was on the 12th day of January, 1867, that the Kiowas made one of the bloodiest raids down the Willie Wally valley in its history. Great had been the suffering, privations and dangers of the settlers before, having now for five years lived in constant dread and alarm. Everything that they owned and held dear on earth was in this valley. Hoping from day to day that the government would help to subdue these ruthless savages, and again bring peace and safety to their once happy homes.

Vain hope! Instead of protection, peace and safety, the darkest cloud of gloom, the most heartrending scenes of rapine, murder and mutilation ever witnessed in any country, fell like a death knell upon us, and left us, like Rachel of old, "weeping for her children because they were not."

It was a lovely bright Sabbath morning, the sun rose in beautiful, glorious splendor, the mocking bird thrilled forth in splendid song, the meadow lark sitting on the old fashioned rail fence, sang her song of praise, the feathered songsters flitting about among the evergreen branches of the live oak, chirped and were glad. And out on a lone tree, a dove was cooing to its mate. Everything in nature, animate or inanimate seemed to be rejoicing. And as I arose this 12th day of January, in my home, I was glad that I was alive. And this was to be an exceptional day. There was to be services at the church in the little village of Forestburg, four miles distant, and I was going to attend, for it was a rare thing indeed in those days when we had opportunities for religious worship. Little did I, or any one, dream that this beautiful and peaceful lovely scene, and happy homes, and everything we held dear on earth, would before the setting of the sun, change from its present peaceful aspect into smoke and ashes, our dearest neighbors and friends mutilated and murdered.

After the services, I went home for dinner with Sam Dennis, one of my old chums. We were sitting and talking together after dinner, when Capt. Hughes came dashing up, his horse in a lather of foam, and asked us if we knew that the Indians were killing everybody and burning everything in Willie Wally valley. And as we looked in that direction we could see the smoke rising all along the valley. No pen can portray, nor have I the language to describe my feelings, for all that was nearest and dearest to me on earth was in that valley. My home, my old mother, my brothers and sisters. Nothing but the restraining hands of Capt. Hughes and my chum's persuasion, prevented me from mounting my horse and at a mad gallop, rush down upon the human fiends who were destroying all my dear ones, but my two friends restrained me, and argued the futility and uselessness of throwing away my life, when it might be so badly needed for the protection of others. As soon however, as we were mounted Capt. Hughes took a course for the mouth of the valley. When we arrived at the lower end, the Indians had already passed on down Clear creek, killing and burning as they went. Just as we rode up on the hill overlooking the valley, we saw Chunky Joe Wilson's house burning just down a little way from us, but we could not see a living thing moving around, so we rode down but still saw no one, and we thought they were all killed or taken prisoners. But as there was only three of us, we did not think it advisable to follow the Indians as we could tell by the signs that there was a large band of them. So we decided to go up the valley on the back trail of the Indians. Imagine if you can my heart burning anxiety, as that devastated trail led in the direction of home and loved ones. We had only gone a little way when we came to where they had killed old man Long and scalped him, passing on up the valley on the trail, we could form some idea of the terrible devastation these savages had wrought, by the vast amount of bed clothes, wearing apparel, buffalo robes, and all manner of household goods that was strewn along their trail. We did not take time to pick up anything, as we were too anxious about our loved ones. As we were loping over the trail, (it was dark or after sundown but the moon was shining almost as bright as day) I saw laying on the trail, a pair of pants, I stopped and picked them up and recognized them and said, "boys they have killed John Leatherwood, these are his pants, see the blood on them." This bloody, silent witness I knew sealed the fate of a noble young man. Every moment increased our anxiety and added speed to our foaming horses. Not a living soul had we seen as we rushed along the bloody trail. When we came in view of my brother's home, my blood seemed to congeal in my veins. Before us was the smouldering heap of a once happy home. Like a madman I drove the spurs into my horse's flanks, and dashed across the valley to my home. And oh! the unspeakable joy, when I found it safe and unmolested, and all my loved ones safe and sound, but not so with some of my neighbors. I found all the settlers gathered here for mutual protection. There we passed a night never to be forgotten so long as life lasts. There were many missing, loved ones of those gathered there.

About 1 o'clock it commenced snowing and got very cold. When daylight appeared we went out to see how many of our friends we could find that had been killed. In a little while we found John Leatherwood, he had only been shot once. They stripped all his clothes off, except his drawers and scalped him. I think it was the next day that we found Arthur Parkhill and Tom Fitzpatrick and his wife, all dead and scalped. The Indians had captured Mr. Fitzpatrick, three children, one infant and two girls of the ages of three and five years. They had not gone far when they took the babe by the heels and dashed its brains out against a tree, and then threw it down in the snow at the root of the tree and left it. This made five the Indians had killed that we buried from my mother's home in one day, besides many others that were killed further down the valley, and Clear creek. The only young lady they captured on this raid was a Miss Carlton, and she succeeded in making her escape that night, slipping off her horse in the high grass while the Indians were having a fight with a small number of settlers. After capturing Miss Carlton about three miles down the valley, they killed old Mr. Long and burned Chunky Joe Wilson's house.

Just below in the valley the next party they killed was old Mr. Manasco and captured Mrs. Shegog and her two children and a negro boy. The Indians then left Clear creek and turned northeast and struck Big Elm near Gainesville, Cook county. And on that Creek Mrs. Shegog escaped from them, but her two children were never heard from any more as far as I know. The little negro froze to death that night. Here the Indians broke up into small parties, and turned back, some going one way and some another, driving everything in the way of a horse before them. This was the bloodiest and most successful raid the Indians ever made in our part of the country. There was not an Indian injured as far as I know, and it accomplished more to stop progress and the settling up of our country than anything else that ever happened before or since for that matter. In fact, everything was at a standstill for a number of years. In our section of the country, however, as time went by, amid privations and hardships, things adjusted themselves again. Here I pause to ask who deserves credit for bringing about this adjustment? It was not the United States soldiers that were stationed along the border, as all the old settlers will agree. It was the brave and hardy pioneers and cowboys who deserve the greatest credit. Men who always stood ready to take their lives in their hands and with the determination to protect their homes and firesides and loved ones, and make homes for them. You may go where you will, east or west, north or south, and you will never again find such a people, fearless, brave to a fault, hospitable and sociable, ready at any and all times to extend a helping hand in time of need. Round them all up and you will find fewer cut backs in the roundup than among any other class of people on earth. When I look back to the past, I am made to rejoice that I can at least in a small degree be numbered with them and helped to blaze the way for future greatness and peace for those who came after us.

PIONEER DAYS IN THE SOUTHWEST

PIONEER DAYS IN THE SOUTHWEST

CHAPTER VI. SHORT STORIES OF PIONEER DAYS EXPERIENCED WHILE A BOY IN PARKER COUNTY, TEXAS. BY JOHN A. HART.

I was born in Madison county, Kentucky in 1850. My mother died when I was two years old, leaving me and a younger brother two weeks old. We were then taken to Indiana, to grandmother's. As young as I was, I remember my mother one time when she gave me a spanking. So the first thing I ever remember was a whipping. The next thing I remember, was grandmother beating apples in a trough to made cider. I suppose I was about three years old then. About that time I remember being at a meeting and hearing the preacher or some one else pray, and the people saying amen. Presently they began shouting and I became frightened; I thought they had got awfully bad hurt and it was not long until I thought all the people had got hurt or were badly scared. My grandmother took me away from the meeting and that is all I remember of Indiana until we started to move. We came down to Texas on steamboat. I remember when we started, that Grandfather Hart shipped his horses and two dogs on the boat. One of the dogs was a bull dog, and would not allow any person to touch him. In order to get him on the boat they circled a beef hide around him and pushed him on. I don't remember anything else until we landed in Texas, and the first thing I remember then was John Hart, a cousin of mine, coming to borrow fire. When he got to the road a big rattlesnake lay before him. The two dogs I spoke of, tackled the snake and killed it, but the snake bit the bull dog, and the dog went to Nolan river about half a mile and lay all day and night in that river. Nolan river heads in Johnson county, near Cleburne, now the county seat. Grandfather died here near the mouth of Nolan river.

I was now five years old and had begun to take notice of things that passed. We left Nolan river and moved up to Parker county. I remember on the way we camped at a place called Hams Hole, where there is a rock at the head of a flat that finally makes a ravine. At the head of the flat is a rock standing up edgeways, leaning over south, about one hundred yards long and is a fine place to camp under for shelter out of the rain or cold weather. We camped here several days and father caught a turtle, the first I ever remember. They cooked it for dinner, and said all kinds of meat was in the turtle. This is the first thing I ever remember eating. I never saw Hams Hole any more until I was about twenty-five years old. I had pictured the rock as being a mile long, fifty feet high, and was fooled to find the rock possessing the above mentioned dimensions.

We moved up in Parker county and there I lived for thirty years. I remember when Weatherford, county seat of Parker, was laid off in town lots. I think this was in 1856. My grandmother built the first hotel in the place. She had a well dug for water and after digging about eight feet, the digger left on Saturday evening not to return until next Monday morning. Grandmother was then camped at the deviate springs, so I slipped off and went to the well and jumped in just for fun. I didn't think of how I would get out but soon found out that I was a prisoner in the well. I hallooed for help but no one came. It grew dark and I was in a fix. Everybody was hunting for me. I went to sleep, after I wore myself out. A young man by the name of Norton came along looked in the well, woke me up and pulled me out. I have always had a dislike for wells ever since. Some time after this the same John Norton that pulled me out of the well pulled me out of a hole of water and saved me from drowning. I went in swimming by myself and got too far out just as he came up and saw me.

I ran away and went to the first school ever taught in the town or country. I took grandmother's big bible, all I could carry and the teacher and pupils made a laughing stock of me. By noon I had sufficient education and quit the school.

Father married again and moved out in the country. A man by the name of Lumslin taught the school and I went to school to him. William Parsons, now living at Clarendon, Texas, was one of my schoolmates at the school. I will mention him hereafter.

Mr. Lumslin gave me a flogging, so this was two whippings I got that I remember, but I might have gotten a hundred more, but I don't remember them. At this school a young man named Emerson played preacher and took Tom Eubanks and myself in the creek to baptize us. He held us under so long we got mad and he liked to have drowned us. He baptized us until we promised to be good, so he said we were good, Christian.

PIONEER DAYS IN THE SOUTHWEST

I never saw Tom Eubanks any more for thirty years. We met accidentally; we had camped for dinner at the same place. During our conversation with each other the subject of baptism came up and Tom related the experience he had at school. By relating the circumstances we recognized each other for the first time. While Emerson was baptising us Parsons hallooed duck 'em good, souse 'em in. After school we went home part of the way together and William Parsons and I had a fight about Emerson baptising me. I don't remember which whipped, but I guess I got licked. If I did that was three lickings I got. William and I fought until we were about sixteen years old. We had more hard fights than any two boys in the county and today there are no better friends in the United States than Will and I. We have fought for an hour at a time and sometimes bloody as a stuck hog. If William had not moved to another state for a while, I expect the matter between us would have been serious, but we are special friends now.

I had a particular hatred for girls and women; they used to run me down to kiss me and I would fight like a demon, but they always outdone me. At one time I slipped up where some girls were holding a protracted meeting. In the play one girl preached and called for mourners, and I slipped up and went to the mourners bench. That started a fight; they run me to the creek and caught me and would have ducked me, but I pulled one in the water and came near drowning her. I then got away and hated girls until I got to the years of puppy love. Then I liked them some better. I well remember the first camp meeting I ever attended. Job Roberson and Beverly Harris, Missionary Baptist preachers, conducted the meeting. All the people in the surrounding country camped on the ground. All denominations took part in the meeting which was held under the brush arbor on Sanchor's creek. Every one seemed to be interested, and one night Rev. Harris was making an exhortation and I thought he was mad and shaking his fist at the people. Some of them were crying and thought they were mad too, and that there was going to be a fight, so I ran to a rock pile and got a load of rocks ready to go into the fight.

Just as I got my arm load of rocks, a woman turned loose, and soon several others. I thought there was a big fight up and threw the rocks down and run with all my might to the wagon, got in and covered up my head and ears. Pretty soon I thought they were all in a big fight, and some of them singing, because the rest were getting a licking. The next morning I made an investigation to see who got licked, but found no scratches nor bruises, and did not ask any questions, but selected several good rocks to throw at the preacher if he hit anyone.

One time father went to mill with an ox cart, a two wheel rig; I heard him coming after dark, and hid in a sumac thicket to scare him, but did not think about scaring the steers. I jumped out and hollooed and away went Lep and Brindle like frightened deers. Dad fell over backwards on the ground, and the steers ran against the cow lot, tore it down, turned over the cart, and spilled the meal. Father grabbed me by the ears, bumped my head against the gate post, and for exercise he practiced with a whip on me.

So Dad, myself, oxen and all got the worst of the fight, but the steers got the most of the fun.

One time father and mother went to see a sick child and left brother George and I to keep house. Father told us not to leave the house, but the dog treed a rabbit over the creek, and we ran over to catch the rabbit which was in a hollow log. I ran my hand in to pull out the rabbit, when a coach whip snake grabbed me by the hand. I thought I was sure a goner, that a snake bite was sure death. I went to the creek, washed my hand and then went to the house and fixed me a good bed to lay down on and die.

After I lain down a while and did not die, I soon forgot all about the snake bite.

One time at school two girls about fourteen years old had some kind of a difference concerning their spelling lesson. They were at the creek at playtime talking over the matter. They had about agreed to drop the matter, and be good friends, or there was talk to that effect, when I happened along and took part in the talk. I talked in such a way that both thought I was their friend and yet I argued the quarrel in such a manner that the girls were soon on the warpath and got into a fight knocking and hairpulling and throwing rocks and sticks and while it was going on, all the rest of the girls came down. Both girls had friends, and other girls got to fighting; I ran off and hid in the brush and watched the whole proceedings. One little girl ran to the school house and told the teacher. When he came, there was a general fight all up and down the creek. When the war was over, their hair was all down, combs, and hairpins down, dresses torn, blackened eyes, mashed up noses and I thought it was the prettiest sight I ever saw. All but four or five girls got a thrashing, while I, the really guilty one, escaped.

I stayed at one place and went to school; a young lady went to school with me. We passed a house on the road where there were about a hundred geese. There was a young woman nearly always after the geese and we called her the goose girl, or at least I did. The girl with me and the goose girl were not good friends. One morning the girl where I stayed did not go to school, but another girl on the road started that morning, and we fell in together. This girl and the goose girl were good friends. But the goose girl did not know that it was the other girl with me, and she paid no attention to us. She was driving about forty geese. I picked up a rock and threw at the geese, not thinking I would hit them, but I hit one goose and broke its leg. The goose girl made at me with a stick. I ran my best, but she was about to catch me when I made a circle and run back to the girl that was with me, and got behind her. The goose girl thought she was the other girl and struck her with the stick, as she was not letting the goose girl strike me; then the fun began; both clinched, and went after each other's hair. Of all the jerking and pulling, I never saw the like. Finally they both got into the water and in the mud, and I thought having a jolly time when the goose girl's brother came and stopped the whole show.

The girl I was with had to go back home for that day. The goose girl and this girl were the best of friends, and ever afterward were the same.

In 1862 in Weatherford some one came in with some friendly Indians. I lived there then; I had not seen any Indians before, and hardly any one else in the town. They were painted and fixed up in real Indian style. After they were gone a few days I proposed to a girl that she and I fix up like Indians and surprise all her people; to this she agreed. I went to a blacksmith shop, and got some red paint. The house was a two story brick, and I got her up stairs to do the work. I had the paint all ready and some big feathers. I painted both her ears and chin red, painted a streak all around the edge of her hair from ear to ear, made a cross on her forehead, and painted her cheeks black. I put the feathers in her hair, and she was a fine squaw. I told her that I would run down and get some more feathers, and then she was to paint me. So I left her standing all fixed up, big Indian heap, and made for other quarters of the town. It was about four years before I could make friends with her again. All the girls I got into trouble at school or anywhere else, I finally got them to make friends.

In blackberry time a crowd of women, children and girls made up a crowd to go to the berry patch and pick berries. In the crowd there were about twenty, all told.

Near the berry patch there had been a negro woman buried several years before. I found out the exact time they were to be there, and I went to the negro grave and hid in the brush. When the crowd all got busy picking, I gave a long lonesome groan. Part of them heard me and began to take notice. One said: "Listen, what was that?" All of them stopped and looked in the direction of the grave, I had now got their attention, and I said with as lonesome and pitiful tone as I could, "Mosey killed me. Mosey killed me; O, Lord have mercy on Mosey." Down went buckets, cups and bonnets; then grabbing of children, and the awfulest stampede I ever saw through brush, berries, over tree tops and the farther they ran the faster they went. Women carrying two-year-old children, could run like a race horse. In about two hours I went to the house and got all the news as to how the old negro came to her death. They heard something at the grave say, "Mosey killed me; O, Lord have mercy on Mosey." I took in all they told me and told them I guessed the negro spirit told the truth. I let on to be very badly scared and finally got off where I could laugh. The next day the women moved, and no one lived there until the close of the war.

George and I went to the cow lot one night and like boys full of projects got to sucking the cows. They had been milked and did not have much milk. I was afraid of the cows. I only pretended to be getting along all right and George said why don't you hunch like a calf? I told him I did not want to. He said you are a coward, just watch me. He gave a big hunch and old Lightfoot give him a side wiper with one of her most busy hind legs, knocked him about ten feet off and broke his nose, it is crooked to this very day.

PIONEER DAYS IN THE SOUTHWEST

During the war some of the old men had got out on a scout to look for Indians. One man got me to stay with his family while he was gone. He had two nice girls. It was awful cold weather. One night I took a half grown cat and went to the girls' bed before they retired and put the cat in one end of the pillows and pinned up the end so the cat could not get out. Soon after dark I went to bed and the girls went also. When they got into bed, the cat commenced to twist and try to get out. This scared the girls almost to death, as they did not know what it was and they ran to their mother, and said something was in the bed. After searching, they found the cat pinned in the pillow and of course laid it on to me. In a day or so a big snow fell and one of the girls got a wash pan full of snow and put it in my bed at the foot. When I got in and stretched out I put both feet in the snow and almost jumped as high as the ridge pole. I thought the bed was on fire so they got even with me on the cat business. After this, the girl that put snow in the bed and myself got into trouble with the girl's daddy, a blind steer and about a hundred panels of fence. It always seemed like that when I got mixed up with the girl's daddy I got the worst of it, but I found it different with their mothers. A boy with the help of the girl can soon gain a mother's sympathy, by telling her we would be good and the word please, and a few other good words, but when dad said no, that settled it. The girl's father had just reset about one hundred panels of rail fence about eight rails high. Before a new rail fence has time to settle it is easily torn down; the least jar will start it to falling. The fence had been completed about ten o'clock, and there was a bunch of cattle going to water, walking down the string of fence. There was about thirty head of cattle in the bunch, among which was a steer with one eye out and could not see good out of the other one. I proposed to the girl that she take a saddle blanket and hide in some bushes, and when the cattle came along, to throw it in front of the blind steer, and see him almost break his neck. No sooner said than she was at her place ready to scare the cattle. As it happened, the girl's father heard me tell her what to do; but we did not know it. The cattle came on down the trail by the side of the fence, and when the blind steer got to the right place, she threw it right in front of him. He gave a snort, ran against a cow and against the fence, knocking down the fence and about a hundred panels of fence fell, making the most terrible racket I ever heard. The girl's father came out all puffed up and was about to give both of us a thrashing. He said that when Brunett, Callie and I got together, the deuce was to pay. Brunett and I had to rebuild the fence and to go at it now. No use to argue the question, we had to build the fence and to go at it at once. It was a very hot day, but it made no difference, we had to go to work just the same.

The first thing we had to do was to move about four hundred rails back out of the way before we could begin rebuilding. By the time we got the rails moved back we begin to get tired of our job, but the hard part was to come yet. We worked until about the middle of the evening and nearly gave completely out. The girl's mother began to beg for us and said her and Callie would help us, but we said no let us alone awhile yet, we were getting along fine. But after while the mother's persuasion overcame the father and he gave up and came and helped us rebuild the fence. The next day was Saturday and Callie and Brunett wanted to go and get some grapes, of course I went along. We went about a mile from home. Came back by the school house where there was to be preaching the next day. The school house was a picket house and a dirt floor. The benches were made out of split logs with legs for them to stand upon. When the benches seasoned awhile the legs sometimes got loose and in moving them the legs sometimes fell out. So, we soon decided to get even with Brunettes dad and for the punishment he had given us on the fence. There were two benches, he generally sit on one or the other and we doctored these two to be sure to get him. We took out two legs from each bench, one from each end, and set them under the benches as props. We dug a hole to set the legs a little slanting so as to make it sure to fall. The next day when I got to the school house several people were there but had not got on these benches. After a while the girls came and next was dad and mother. To our surprise dad did not sit down on the benches we had fixed for him but mother did. The girls were rather scared but I was tickled almost to death, yet I was disappointed. Dad did not get on either bench. After while my grandmother came in and sit down on the bench by the girl's mother. I was so full of laugh I could hardly live. I was actually in misery and the girls scared nearly to death. Grandmother began to fan with a turkey wing and seemed comfortable. She was a big fat old lady and I knew when the time came great would be the fall. At last the show came off. It was like a railroad wreck. About this time down came the bench. Grandmother and all the rest spilled on the floor together. I was so tickled I had to leave the house and did not see any more of the performance. Yet, I had not got the one I was really after. In a few days we tried another trick to catch dad and got the girl's mother again. The father had a foot log over the creek, this made a narrow way to cross the creek, even when the creek was down. Each end of the log lay flat down on the bank of the creek. We decided to undermine one end of the log so when dad got on the log it would break through and fall in. We dug under the end of the log and left it ready for the job. The girl's mother went to look after some young goslings and had to cross the log. She got out on the log and it gave way and down she went into the water head and all. I was not there but the girls said it was a sight.

On Burges creek Joe Kaughman gave a social party, old and young were invited. There were lots of people gathered at Uncle Joe Kaughman's. He had a large family of boys and girls. They were old time Arkansans, but had been in Texas a long time. Everything that was good to eat was put on the table and more boiled goose eggs than I ever saw. There were several tables set and everyone seemed to like goose eggs. It was something the people never saw before, was goose eggs on the table at a supper. There was a young man from town come out somewhat stuck up. He took a fancy to one of our country girls and some one gave him an introduction to her. She was rather a retiring disposition, some people would say bashful. They ate supper at the third table. All the goose egg shells had been cleared from the tables and was all in a pile behind the stove. I got a big pan and got all the goose egg shells and put them all around under their chairs. I never saw the like of shells around one couple. The young man and girl both took a goose egg for a rarity. After a while a little boy discovered the egg shells and hallooed out "just look at the goose shells." Of course this drew the attention of everybody. They were the worst looking couple I ever saw, and anyone can imagine how the people looked and acted. Uncle Joe tried to find out who put the shells there. There was more than a dozen saw, yet no one knew. Uncle Joe didn't care, he was only trying to smooth things over; for Uncle Joe enjoyed the joke as well as anybody.

Down on Clear Fork creek a dog came to a man's house. The dog was no account in the world. The man told his daughter and I that if we would take the dog down on the creek and hang him he would give us a party. Of course the old man did not think we would do it. But the girl and I finally concluded for the sake of the party, to do it. I had hard work to get the girl to agree to help hang the dog. The girl's father had left the house when we finally concluded to hang the dog. We got a rope and led the dog to the creek, put the rope over a limb, and drew him up and tied the rope to another limb and ran off, as we did not want to see the dog die. In about half an hour we went back, and the dog was dead, so we took him down and went to the house. The girl's mother came very near giving us a thrashing, but the father was to blame. He said he did not think we would do it, yet we should have the party. So everything was arranged for the party next night but in the morning the dog was back at the house ready for his breakfast. The old scoundrel had come back to life. It seemed as if everyone on the place was glad the dog came back. The old folks gave us our party, but we both agreed not to hang a dog that could not be killed. But the same day we drowned a big hog too quick and easy to realize what we had done. The hog lot gate got open, and one of the hogs got out and we hemmed him up and caught him across the creek. He was a large hog and heavy to handle. We had to take him by the hind legs and push him to the lot. The nearest way was through a hole of water, about knee deep; but we did not care for the water. So we pushed him in and shoved him across, his head under the water. When we got him to the opposite bank, we had all kinds of a time to pull him out, and when we did he was as dead as a hammer. We expected to get a scolding and dreaded to go to the house; but when we did, and explained the matter, we were not blamed, while, of course everyone hated it.

I have not given all my little funny pranks, and while I had my fun, I had some hardships, and if you will read the Pioneer History, you will learn what a frontier boy in Texas had to experience. By reading this book, you will learn a great deal that is profitable.

PIONEER DAYS IN THE SOUTHWEST

PIONEER DAYS IN THE SOUTHWEST

CHAPTER VII. HISTORY OF PIONEER DAYS IN TEXAS AND OKLAHOMA. BY JOHN A. HART.

I WILL endeavor to write a short history of pioneer days, incidents that happened from 1855 to 1879, of my own experience and the experiences of men, women and children of old pioneer days. This is no dream or idle tale, but in every sense a true history. My object in writing this book is that there are thousands of persons living who are ignorant of the hardships, trials and difficulties that people of old pioneer days had to contend with. A history of this kind will be instructive to the rising generation and interesting to the old people who yet live and who experienced the old pioneer days. I have sometimes seen old pioneers meet, and in relating their past experiences have noticed people stand and listen in wonder and amazement, thinking and wondering if these accounts were true.

I have no doubt, a great many imagine that people of pioneer days were toughs and outlaws, but such is not the case. While I was only a small boy when Parker county was first settled; only about thirty families in the county. I remember that nearly all heads of families were Christians at that time and I am of the opinion that Parker county is only a sample of the people of pioneer days in Texas, generally. I will admit there was a great many tough people in one way, yet, when it came to a point to protect life and property they were willing and ready to give a helping hand, and I am satisfied and can safely say of the settlers of pioneer days as a general rule, a more kind hearted and generous people never lived. I know all of the old people of that day and they will agree with me on this point without hesitation.

I was five years old when Weatherford, Parker county, Texas, was laid off in town lots. John Prince taught the first school that was ever taught in the county at Weatherford. I was too small to go, or at least my parents thought so. I differed, so one morning I shouldered the old family Bible and ran away and went to school. The Bible was all I could carry. By twelve o'clock I decided I had education enough and quit school.

All houses were built out of rough logs which were snaked out one log at a time. In this way enough logs could be dragged out to build a house, joist, rib, weight poles and all, in a week. Most all houses were made of round logs, not hewed, from ten to twelve rounds high, two foot boards were laid on rib poles, and weight poles laid on the boards to hold them down. Stick and dirt chimneys with rock back and jams, a door of clapboards chinked and daubed with mud and a dirt or puncheon floor and the house was completed and good enough for a king.

Wood hauling with steers was very common. Tree tops would be snaked from one-half to a mile and sometimes further.

Furniture was generally very simple. Sometimes a one-legged bedstead, and sometimes four. In a four-legged bedstead an augur and chisel were used to mortise places for the sides and end pieces. Holes were bored in the side and end pieces about every eight inches apart and a raw hide rope run through the holes, while green which made the bottom or a place on which to put the bed. A one-legged bedstead was different. A hole was bored into the wall and the side and end pieces put in the holes and the one leg built on the fashion of the four legs. Slats of boards for the bottom laid across and you had a bedstead fine enough to invite anyone to sleep on.

The dining table was built on the same fashion as the four-legged bedstead. Boards were shaved out with a drawing knife for the top and when visitors came, a tablecloth was used, made of domestic, woven by the housekeeper herself.

Some were lucky enough to get hold of a few chairs and had stools enough to make out the set, some had all stools or benches. A tree was cut and split open and faced with a broad axe, four holes was bored and legs put in and the stool or bench was completed.

The cooking utensils consisted of a three-legged skillet, oven, dinner pot, tea kettle, a big iron shovel and a pair of pot hooks.

I very well remember the battling stick on wash day. Oh! my, how tired I would get on wash day beating and pounding the clothes while mother rubbed them with her hands. The clothes would be taken out of the tub and laid on a bench or block made for that purpose and beat with one hand and turn the clothes with the other. The battling stick was in the shape of a paddle, only heavier. This was before the scrub board came in fashion, and then we thought a great improvement had been made when the scrub board came in.

PIONEER DAYS IN THE SOUTHWEST

Candles to give light were all made by hand there were two ways to make them, one was to get as many sticks as was needed and tie strings to them, usually about half dozen to the stock for wicks and dip them in a pot of warm tallow, lift them and dip them in a pot or bucket of cold water and back into the tallow. I have seen several hundred made at a time this way. Some had molds, drop the wicks in the molds and pour the tallow in the molds. There were two kinds of candle sticks used when a light was needed, one was made out of tin and the other a square block with a hole bored in the center to set the tallow candle in.

All the soap was made by dripping lye from an ash hopper. To make an ash hopper, first drive four forks in a square, put side and end pieces in the forks, this made the frame, put a trough under the frame and stand boards in the trough and the top end of the board against the frame, fill up with ashes and pour water on the ashes for about two days and the lye will start to drip and an oven or skillet is placed under the end of the trough to catch the lye. There are thousands of people today who do not know how lye soap and tallow candles were made.

Oxen were used in farming to a great extent. In breaking prairie from two to four yoke of oxen were hitched to a plow. In old ground from one to two yoke were used in plowing corn and garden stuff. A home-made turn plow was used to break land; a bull tongue to cultivate; and eye hoe that weighed about one-half to a pound was used to cut weeds and grass. Corn was dropped by hand and covered with a hoe.

Wheat was threshed out by hand, or tramped out by horses. If threshed by hand the heads were beaten out with a pole; if by horses, the wheat was spread out in a yard prepared for the purpose and from six to ten horses were rode round and round over the wheat, men stationed in the inside kept the wheat stirred up, if the wind blew, the wheat was poured out of a bucket to blow the straw and chaff out, if there was no wind two men took a sheet and fanned, while one poured the wheat out of a bucket.

In traveling to church some would walk from one to three miles, others ride on horseback and a great many went in ox wagons. All the wagons were tar pole or wooden axle. The men were generally armed either with guns or pistols, while at church the guns were stacked by the side of the house, revolvers never taken off.

At camp meetings an arbor was built for services and small arbors for the campers. When the campers all got to the camp ground the steers would all be belled and hobbled out. Everybody was welcome and at home. A beef would be killed every day and divided with all the campers, the beef did not cost any one any thing, people who lived twenty miles away got beef, no one thought of charging for it and the people on the ground were like one family of kinfolks. Everyone wore homemade clothing, shoes and hats. I have seen some leather breeches and coats. They were only buck skin but called leather breeches. Some had caps of deer skins with the hair side out. All had their guns at church, with their shot, pouch and powder horn. In the pouch would be a bar of lead, bullet-molds and a rag for patching, and if caps were used, a box of caps, but if a flint lock, several flints.

No one ever thought of charging a stranger or traveler for lodging. It was an insult to offer to pay for a night's lodging. Stock hunters could travel all over Texas and never be out one cent.

Milling was sometimes done with an ox cart, a yoke of oxen to the four wheels of a wagon and corn piled upon the cart for four or five families.

A wedding was generally public, everybody invited and all went and a grand charivari followed. It was the custom to have a grand time Christmas, and if people were to celebrate Christmas now as they did then, they would be considered regular outlaws and all be arrested.

Rope hobbles, bridle reins, clothes lines, bed cords, were nearly all made out of raw hide. Some hair rope was made out of hair by cutting the bush of cattle's tails off and twist the strands into a rope. People were considered quality when they could have a pair of hair bridle reins.

They always wanted to know when it was going to rain or when the sun or moon would rise and set and they would pay ten cents for an almanac. I remember when matches sold for ten cents a box which contained twenty-five matches. Only travelers or freighters could afford to use them. Matches were not used only in extreme cases. Many times I have walked a mile to borrow fire. Everybody kept a piece of punk and a flint rock to strike fire, and by placing the flint on the punk and striking it with a pocket knife would produce fire. Sometimes we would take raw cotton, place it on a skillet lid and sprinkle powder over the cotton, take a case knife and strike the lid, knock fire out of the lid and catch the powder, and we had fire.

Wheat was cut with a cradle by hand, and a good cradler could cut about two acres a day. A good binder could keep up with a cradler, but generally an acre and a half was a day's work.

Three or four families owned a sorghum mill. The rollers were made of live oak and the cogs were all wood. The owners of the mill assisted each other in making sorghum. We did not know sorghum by that name then, just plain molasses. A single horse or steer turned the mill. It usually took a driver to keep a steer moving. Each person would make from one to three barrels of molasses. When molasses making was over the youngsters had a candy pulling, old people and all would take part.

In case of sickness or distress every person showed a willing hand in such cases. If any person got behind with his crop, wheat-cutting or anything else all turned out to his assistance.

Work steers were like all other animals, they had different temperaments and different dispositions and some steers if treated kindly were easy to get along with, while others didn't know or appreciate kind treatment. I suppose a great many old timers who have hauled water on a slide or lizzard have had something to remember all the days of their life, provided you used a single steer for a team.

A young man told the experience he had hauling water on a lizzard with a single steer. A lizzard, as we called it, was a forked tree cut down and the fork of the tree was the sled. Cross piece put on to set the barrel on and four standards to hold the barrel on the lizzard, put a half yoke on a steer and hitched to the lizzard and you were ready to haul water.

The young man hitched the steer to the lizzard and with the assistance of his sister drove half a mile for water. It was a warm day and tiresome, and by the time the barrel was filled they were both very tired. They drove about half way home when a heel fly struck the steer on the heels. A steer is very sensitive about a heel fly when it tackles a steer's heels so it was good-bye steer, lizzard, water barrel, water and all. No use to try to stop a steer when a heel fly gets after him. The water all gone, the barrel at one place, the lizzard at another and the old steer down in the creek bottom in a thicket, looking very innocent.

My grandmother was a great hand to make soap. She had filled the ash hopper with ashes and poured water on them until the lye had begun to drip and had run out of water. Grandma and one of the girls hitched old Nig to the lizzard and were off for a barrel of water. Grandma carried the bucket and the girl drove old Nig. Grandmother was a large, fat old lady and it being a warm day made the trip hard on her. Old Nig brought the water back to the yard gate all right. Grandmother went to open the gate so the girl could drive Nig into the yard with the barrel of water; about the time the gate was opened a heel fly just to be friendly with Nig, visited his heels, away went old Nig and tore down the gate, run against a stump, upset the barrel of water, run against the ash hopper and tore it down as flat as a pancake. Old Nig backed up in the shade of the smoke house and looked as though he had made a great victory.

All you old timers know what a boy will do in a case like this, who has had no hand in the promenade. I was sitting on the yard fence and did not need a spy glass to see the whole performance.

In plowing corn with a steer, you might work him for a week all right, but before you could take time to think he would run away and tear down the width of two or three rows of corn. It seemed the steer had all the fun and the driver all the trouble. Nothing ever pleases a boy more than to see a steer and a woman get in a mixup together.

Some people may think a steer has no sense, but this is a mistake. If a steer was properly broken to work and kindly treated you scarcely ever had any trouble, but you always had to be careful in heel fly time, because a poor cow or steer, so poor they could hardly walk would run from a heel fly when nothing else could hardly get them to move. We had a yoke of steers that would feed all night and about day light would hide in a thicket and hardly move all day nor rattle their bell. Many a time I have gone out in the yard about daylight to listen for the bell to get the course and every duck on the place would turn loose with all the quackism in their power, and the guineas would turn loose and the turkey gobbler would gobble. There is no use to try to quiet a bunch of ducks, guineas or gobblers when you are trying to listen for a steer bell for the more you try to make them stop the more fuss they would make. I said then if ever I owned a home and a family, I never would allow a duck or guinea on the place, and I have made my word good.

One time Judge Embree concluded to go to Weatherford to mill, he hitched a yoke of oxen to the front wheels of the wagon and put on two or three sacks of corn; Mrs. Embree went with him to market with some eggs and chickens, she had two dozen eggs and four chickens. The got about half way to town when the heel flies got after the oxen, they left the road- on double quick time and the eggs, chickens, corn, Mrs. Embree, the Judge, the cart and steers were all scattered. The Judge got the cart and oxen and gathered up the corn but Mrs. Embree lost the eggs and one chicken in the runaway scrape.

When I was a small boy I drove from four to six yoke of oxen to a freight wagon. In those days ox teams were all the go, but now that the oxen's days of labor are over I cannot help but have a kind feeling toward the oxen. At the word of command I have seen a yoke of oxen at the wheel hold back and stall five yoke of oxen.

During the war we fed prickly pear leaves to cattle; we would take a wagon and haul the pear leaves home, some times as many as eight or ten loads. The prickly pear has long thorns and thousands and thousands of small stickers on the leaf, We would haul wood and brush and build a fire and singe the thorns and stickers off the pear leaves and then feed them to the cattle. I have had the stickers all over me and in my clothes by the hundreds but we had to do this to keep the oxen and milk cows alive during the winter. The live oak was an evergreen tree, and they were sometimes cut down so the cattle could eat the leaves.

Squirrel and cat hides were dressed by the boys and girls for shoe strings and whang leather; raw hide was used to cover saddles, and sheep skins were used for saddle blankets; some used a hair rope stretched around the bed when they camped out, to keep snakes and tarantulas away, as they will not cross a hair rope.

Sumac leaves and black jack bark were used in tanning leather; the leaves were gathered while green and the bark peeled off while the sap was up, and could then be stored away and used any time of the year. The sumac leaves were used for shoe leather and the bark for sole leather. When the hide was ready to tan, the leaves or bark was boiled, the hide was then placed in a trough, the pulp was placed between the folds of the hide and the liquor poured over it, this was repeated every three or four weeks, sometimes not so often.

No one thought of going in debt, they paid cash or its equivalent, or did without; if someone would have offered to take a mortgage the people would not have known what it meant. Borrowing and loaning was very common, everybody's oxen and wagon, or anything else was to loan if anyone wanted to borrow it; and whenever anyone looked for their stock they looked for their neighbor's stock at the same time.

At a house raising the women came too, and assisted with the preparations for the fine dinners prepared on such occasions. At a quilting there were enough women come to do the cooking, the men would come and enjoy themselves pitching horse shoes, running foot races, jumping, telling yarns or anything else for pastime.

Sometime I almost wished I were a girl so I could have a good time, but I had no sister large enough to work so I had to churn, wash dishes, use the battling stick on wash days, make bats for quilts, quilt and hand the thread through the harness of the loom and then I was glad that I was not a girl so I could get out of such work. I never could believe that I was cut out for a boy and a girl too and such work now would hurt my feelings terribly, but there were lots of girls that did a boy's and girl's work too, and lots of women that did men's work in war times. After the war a family by the name of Williams came out from Arkansas and I heard the woman relate her experience, which I will give to the best of my recollection, I don't remember the part of the state she lived in, it has been thirty-five years ago and I don't remember all she said. Her husband was in the southern army, left her with four small children and not very much to live on, she had a little peg-horn yoke of oxen, two cows, an old run down wagon and four or five little knotty hogs when her husband left home to go to the army. The next year was a trying one to the people in that part of the country. She heard one day that the federals were coming in that direction and to remain at home meant destruction to all she had, so she resolved to hide in the brakes and mountains and move everything she could. Her old wagon was almost gone to staves, she took some soap and greased the wagon to keep it from making so much noise, drove wedges under the tires and poured water on the wheels to soak them up, she put part of an old cow hide in the creek to get material to make ropes, then gave the children instructions and went for the oxen. After hunting for several hours she found them, but she was very nearly worn out, she had to keep going though and make preparations to load up; her children wanted supper but there was no time to cook it. It was dark when she got her household goods on the wagon; and the oxen had not been worked for some time and were very unruly but she was determined to save what she had at all hazards. The older children walked and carried the two younger for a short way until she got the oxen so she could manage them. There was a scope of country for several miles that was brush covered brakes and in this she intended to hide. By hard work she managed to get to where she wanted to go; she tied the oxen to a tree, got supper for the children and put them to sleep, then stood guard the balance of the night as the woods were full of wild animals. The next morning she built a scaffold to put her meal and meat on, she had two small loads of corn to haul; next she hauled a load of rails, built a pen for the corn and got a load of corn that day, and from the cow hide she put in the creek, she cut rope and staked one of the oxen out on the grass and the other one stayed without tying; by this time she was almost afraid to go for the remainder of her corn, but would save the cows if possible; she found them all right but had a hard time getting them into camp. There was only one calf and by giving the cows a few shucks and a spoonful of salt she caught the calf, tied it to a stake and that kept the cows close; she had a few chickens but did not take them. She thought she would go to the nearest house and see what she could learn about the Yankees. When she got in sight of the house the yard was full of them so she ran back to camp and did not go away again for a week, then she went to another house and it was burned up and no one there. The next morning about sun rise, she heard the sound of guns and lots of them and she was in hearing of guns and cannons for two days, but she stayed this out for ten days longer, then went out prospecting and met a boy who told her all the news. Nearly every house was robbed of everything of value; some had a little left that they had hid out, he thought that the Yankees had gone but he didn't know where. She then went back to her own house, the corn and chickens were gone and the yard fence burned up.

She then went back home and was so glad she had a little left. Then one of the children took sick and died. She made the coffin and dug the grave herself, and with the help of another woman buried it. She divided her corn with her neighbors and soon they were all in a hard shape. One steer died; one cow killed by lightning. The soldiers got all the hogs but one and from this hog she raised meat. She says she can't tell how she got along, but she got along somehow. One evening she was sitting on the doorstep thinking what to do next, she looked up and lo and behold, she saw Mr. Williams. "I did not know him until he spoke, and the first words he said were, 'Sally, war is over.'" She did not know any more, for she had fainted. I have an idea Aunt Sally did have a hard time. For a little while she was too much affected to talk further.

Later on when questioned about hiding out she said it was all through fear that made her do it. She was afraid to stay in the woods but more afraid of the Yankees and her thoughts were all centered on getting out of the way of them and saving what she had. It rained often on them, and sometimes they were drenching wet, and for several days at a time she couldn't leave the children to go anywhere, on account of the rain. At night she tied the calf to the wagon wheel to keep varmints from killing it, and then she had to keep a fire burning all night to keep them away. The owls and the night hawks made it very unpleasant. She said she fared as well as lots of women.

Some had their houses burned up or everything taken from them they had to eat. The Jay Hawkers came through occasionally, and they were worse than the regular soldiers. On one occasion, about seventy-five men camped at her house, the captain was very inquisitive and asked many questions. She told them her husband was in the southern army, but she was satisfied they knew this without asking her. They did not molest her, did not even burn a rail. They stayed three days and she washed handkerchiefs, socks, a few shirts and sewed buttons on their clothes and they paid her more than she asked. When they left they gave her some scraps of meat, some beans and a lot of old clothing she picked up and used and glad to get them. She learned afterwards that they were Captain Quantrell's Independent company and that a lot of them were killed on the Missouri side. It is impossible for me to relate all this old lady told. She was a good Christian and passed out of this world of sorrow about twenty years ago.

During the war Mat Heffington, who lived about five miles south of Weatherford in Parker county, went to Dallas county and bought a negro. Heffington wanted the negro to look after the stock. The farm was small and nothing but corn raised on it. The negro had but little to do but look after the stock. His name was Jack. Several months after Heffington bought Jack he sent him to the range to look after the cattle, but several hours after Jack left some one notified Mr. Heffington that Jack was running away, on the road back to Dallas. Heffington overtook him and brought him back, but did not punish him. He did very well until the next spring and then lay out in the neighborhood and prowled around people's houses for something to eat. He was captured several times in summer, but would hide out as often as he was caught; whipping did no good. At last Heffington put a big pair of iron hobbles on him and put him to chopping poles, but he finally ran away and was gone about eight months. Winter time set in, and he came in and said he would be a good negro. But spring time came and he was gone again. He would waylay school children and take their dinner from them, and sometimes cut all the clothing off the girls. He cut the clothing off one young lady and she requested when he was captured that she have the privilege of whipping him. He was soon captured and tied to a tree and the lady sent for. She preferred a hand saw to a whip. Jack was stripped to the waist and left to the mercy of the lady. When she got through he was a solid blister, she salted and peppered his back and turned him loose. He ran to the creek and lay in the water for several hours, came out and begged to be taken back, but soon was gone again. When the war ended he was still out, and in August was killed, and found several weeks after, about three-fourths of a mile from a house, in an elm thicket. About one year after, there was a party at the house near where he was killed and a young man offered the young lady of the house, who was about thirteen years old, his horse, if she would go down to the thicket and bring Jack's head. It was awfully dark, but the young lady got up without a word, put on her bonnet, opened the gate and disappeared in the darkness, in about half an hour she came in with Jack's head rolled up in her apron, and laid it down before the young man, there was all kinds of talk when she laid the head down. He gave her the horse and paid her to carry the head back the next morning. There is not one woman or girl out of a thousand of this day, who would go in the night and do such a thing as this girl did. And what woman or girl would pick up a log chain and go between a yoke of steers and hook the chain in the staple of the yoke, or go between the long horned steers' heads and put the tongue of the wagon in the ring, or put a steer's leg back if he got out of the traces. No, you would run and squeal like a panther if he licked out his tongue begging for a little salt. The gate post would not be high enough for you to climb up on.

Girls are mere babies now to what they were forty years ago. I knew one girl eighteen years old who took a gun and killed a rattle snake six feet and five inches long. What would you girls do today? You would run off and leave the snake to get away, of course.

Women of old pioneer days, could and would do anything that was a necessity, and do it willingly and cheerfully. They would even take a gun and stand guard in time of danger of Indians and if necessary would fight in defense of themselves and families, and some of them have been known to do so and come out victorious.

In the year 1870 or 1871, about fifteen miles above a town, I think it was Waco, or at least that is my recollection, but it might have been another town on the Brazos river. There was a ferry boat on the river that ferried when the river was too high to ford. A man with his wife and child about two years old, drove on the boat in a two horse buggy. The Brazos river was nearly bank full. The man took the horses from the buggy, or rather unhitched the traces, but the woman and child remained in the buggy. When the boat was about one-third of the way across, the boat dipped water and sank. The ferryman and the man swam out, the horses drowned, but the buggy floated down the river with the woman and child in it. A runner was sent to the town and men, women and children gathered at the river. Nearly all the town was out. When the woman and child came in sight, all the people went wild. Some men offered fifty, some a hundred and some five hundred dollars for any person or persons who would save the woman and child. There were two cowboys sitting on their horses taking in the whole situation. One of them proposed to the other if he would save the child, he would save the woman. The boys leaped from their horses, stripped, and in they went.

They had to go down the river nearly a mile before they caught up with the buggy. All the people followed down to see the boys bring the woman and child out. The boys climbed up on the buggy by the woman. They told her that they would bring them out safe if she would do as they said. One took the child and leaped into the river. The other told the woman that if she took hold of him any place that he intended to drown her, but if she would keep her hands off of him he would take her out all safe. He had her take down her hair and climb on his back. He took a twist of her hair and put it in his mouth, and in he went. The woman seemed to realize what he had told her and had all the faith that he would, take her out. The boy with the child landed all right and great shouts went up from the people. The woman on the boy's back laughed, but soon she was landed safely herself. She took the boys by the hand and cried, and thanked them and they cried too, and there was general rejoicing.

The two cowboys would not take the rewards offered. It looked cowardly to them to take a reward. I have heard men say since then that the cowboys of that day were a set of outlaws, but I will say to all such men that the pioneer cowboys fought more hard battles to protect Texas than all the soldiers ever did. The cowboy would fight to his death for the defense of a woman's life or character. The man of today who calls the old pioneer cowboys outlaws, would offer a reward of a few dollars to save a drowning woman, but too cowardly to risk his life as the two brave cowboys did to save the woman and child.

I have had some experience in the cowboy life and when they got on the trail of Indians that had killed and scalped men, women and children, they were like all citizens who followed the Indians, when they would run down their horses and fail to catch them, would curse and cry because they could not overtake them.

In 1874, on my last buffalo hunt, on my way back home at Fort Grimm, General Philip Sheridan had whipped the Indians, killed their horses and the Indians had to walk in the best they could. The Indians had a war dance. One old squaw had a white woman's scalp on a pole, the Indians dancing around it. A cowboy pulled his revolver and was going to shoot the squaw and swore he would kill her, but was prevented by the soldiers. The scalp was taken away from the squaw and given the cowboy. No cowboy or citizen could stand and see the Indians dance around a white woman's scalp. There would have been a hot time just then if the cowboy had been left alone.

In case of sickness or some one wounded, no one would ride farther and faster for a doctor than a cowboy. All boys in middle and west Texas were more or less cowboys, as all the men were more or less interested in the stock business, and all those boys and men became the best citizens of the country; and yet some people of today will say they were regular outlaws.

The women of Texas were not as easily excited, or in other words, in case of danger would not lose their presence of mind as easily as some might imagine. Some of them have been known to count the Indians and sometimes the horses when the Indians were only a short distance from them

Mrs. Sarah West, who was Miss Sarah Powell before marriage, counted the seven Indians that were killed on Robertson's creek, one squaw in the bunch. On Squaw creek in Hood county, Mrs. West had just finished washing, about on hundred yards from the house, and went to the door and got a drink of water and turned around and saw the Indians carrying off her clothes. While they were gathering them up she counted them. She stayed in the door until they were through gathering her clothes.

Near the Comanche peak two girls, one sixteen and the other six years old, were gathering grapes on the side of a rock ridge, with prairie on north side. The girls heard horses and on investigating discovered a party of Indians driving a bunch of horses. They hid in the thicket and the older girl counted fourteen Indians driving seventeen head of horses. The Indians passed within about seventy-five yards of them. The little girl said "there was a great big bunch of Indians with a big black bonnet on, riding great big horses." They were all great big Indians driving a big bunch of horses. Two miles from there another woman stood in the yard and counted the same Indians; there were fourteen. Further on a young man and a girl counted fourteen. Mrs. A. P. Parsons and Mrs. Eppie Gardner counted the Indians while they were taking the horses out of the field. This was the same bunch of Indians that killed the settlers. Mrs. Keener counted nine Indians that passed between the house and field near the Littlefield bend on Sandy creek in Palo Pinto county. Three women counted the same nine Indians at the sorghum mill, eating cane and skimmings from an old bucket.

It was a noted fact that when Indians passed a house, if a woman was at the gate, door or in the yard and kept quiet, showing no signs of fear, she was rarely molested, but the least sign of fear or alarm and they would go after her scalp. Indians never killed an old person, a cripple or a deformed person, if they knew it. Women were close observers in Indian times. They learned that the whoop of the horn owl and that of the Indian were different. The voice of a horn owl will not echo, the Indian's would. They would pass through a neighborhood in the night, string out abreast for several miles hunting horses, whoop like an owl and in this way could locate each other.

Every person in Indian times kept two or more dogs and in war times the women learned when the Indians were in, where the most danger was, by the bark of the dogs; all the dogs in the neighborhood would bark. If the Indians were far off the dogs would go out in the yard to bark, if they were near the place, the dogs would come inside near the door and probably bark very little, but act very restless. It seemed that an Indian was a mesmerizer or something for the dogs.

In 1873 near Pecan bayou, a woman heard a calf bawling like something had it, she thought the bawling was the voice of an Indian, and it might be to decoy her out. Next morning when she went out she found Indian tracks at the lot and all around the place. In that same community another woman was alone and some one called, "Hello!" She noticed the hello was given in a downward voice and she had noticed that all calls at the gate by white people was with a rising voice. The hello was repeated several times with the downward voice. The woman managed to see out, and in the fence corner near the gate sat an Indian. She was armed and determined to defend herself and children, but not shoot unless it was necessary. The Indian finally left as his call was not answered and if answered might be with a gun. So you see people were put to their wits end to outdo the savages and by doing so save their own lives and the lives of their families. Had the woman gone to the cow lot or answered the Indian's call by going to the door, both woman and family would have been murdered.

PIONEER DAYS IN THE SOUTHWEST

Ever since I can remember until the year of 1868 or 1869, every two or three years we had in the fall of the year the traveling grasshopper. They came through Kansas, Oklahoma and Texas. I have seen them when they first begun to light, only a few, and in thirty minutes they would be falling all over the face of the earth. They would be so thick between the sun and the earth that they would cause a shadow on the earth just like a cloud over the sun. I have seen them so thick on the trees in the timber, one could scarcely see the bark of the trees, and I believe I would be safe in saying I have seen as many as a bushel on one tree. Where I lived was in a thick timbered country and the whole country was covered with grasshoppers alike. While they did not damage the grass to a great extent, they ate everything else. I have known them to ruin a nice garden, clean it all up in a half day's time, even the dry fodder on corn. Not long ago I related the old grasshopper times to a man who was surprised to hear such of the old time grasshopper and he seemed to think that such a thing was impossible, but all old timers up to 1869 will testify to what I have said concerning the old time grasshopper in Texas.

During the days of the buffalo, Aaron Hart and some others, while on a buffalo hunt between Little and Big Wichita rivers, found the skeleton of a woman tied up in a sheet and hung up in a mesquite tree about twelve feet from the ground. They supposed from the appearance of everything that she had been there two or three years. The sheet was rotten and ready to fall, the clothing was all cut up by mice but her hair was yet done up, her shoes fairly well preserved and also a comb and some other articles. They took the body down and gave it as respectable a burial as they could. There was no settlement nearer than Jacksborough at that time. No one could guess why she came to be buried in that style and how she came to be in that part of the country, unless she had been carried there by the Indians and then it was not like Indians to bury in that manner. Whether she died a natural death or was killed, no one will ever know.

In 1868 a party of white men were camped near where Breckenridge now stands. After dark they heard Indians pass about a quarter of a mile away. They heard a child crying which they supposed was two or three years old. Afterward they learned a child had been carried off by Indians north and west of Gainesville.

PIONEER DAYS IN THE SOUTHWEST

The time the two savages were killed, of which you see an account in J. D. Newberry's chapter, three children were carried off, two boys and a girl. One of the settlers was plowing with a gray horse, an Indian unhitched him from the plow, tied a long rope around the horse's neck and made a halter; put the children on the horse then jumped on and put the coil of rope around his own neck. The next day the horse came home dragging the rope. The horse ran away with the Indian, threw him and pulled his head off. This story was related by the girl two years later, when she came back home. She said the old squaw carried the Indian's head in her lap. Two days later they tied his body on a horse, they dug a place to bury him and put him in the grave and all got around to cry. They never cried until then. They tried to make her cry but she kicked dirt on them. They slapped and pinched her but she would not cry. The boys cried after being pinched and slapped. The three children tried to run away. Once they got about a mile away when they were overtaken. The Indians stood them on the edge of a bluff and pushed them off backwards. They were taken back and the ball of the girl's big toe cut off and poisoned. One boy had a gash cut in his nose and the other on the head, and both poisoned. They were gone two years. Their wounds were all running sores when they came back. The girl said in winter, when it was freezing cold, they would be thrown in the water and then wrapped in buffalo robes where they kept warm all day.

I live now about eighteen miles from the battlefield where General George A. Custer and the Cheyenne Indians had their fight. The battle ground is in Roger Mills county, Oklahoma, about two miles from the county seat. I have not been on the battle ground, but I am informed that the bones of the Indians and horses can be seen all over the ground. Six miles from here, at Hammon reservation, a lot of these Cheyenne Indians now live in their tepees, that fought General Custer. It was about the last battle the Indians had with the whites.

While the Indians were on the war path they used poisoned arrows almost exclusively while fighting their enemies. I learn from the Indians the way they prepared this poison was to take a rattle snake's head and boil it and hold the arrow spikes in the steam of the boiling snake's head and the poison settled on the spikes. It was sure death when the point of an arrow struck and made a wound.

PIONEER DAYS IN THE SOUTHWEST

The Cheyenne and Arapaho Indians are the filthiest people on earth about their eating, at least all those were who are here on the Washita. They will pull dead cattle out of the mud that have been dead for a week, skin and eat them. They even strip the entrails and eat them. Mr. Charley Wagoner gave one two dead shoats that had been dead two days. He carried them off to eat. They will eat a chicken that died of some disease, or a dead horse if it has not been dead more than a week in summer time. The white people here can trade them a cripple or diseased hog, cow or horse to eat and some of them trade with them.

The Indians used shields in war times to stay off the bullets or anything fired from the enemy. The shields were made of raw hide stretched over a hoop while green. This hung around their necks and always turned toward the enemy and were used to great advantage while in a fight.

In 1876 I had the curiosity to attend an Indian cry. When an Indian died, a time was set for the cry. Not a tear was shed at the time of the death and burial. The day set for the cry, there is a feast first given by the relatives. When the feast is over, a signal is given when all march to the grave of the dead, all circle around, the women and children in front. Then the cry begins, and of all the squealing and howling I ever heard in my life. There is no foolishness about their crying. They really weep. The Comanches, Kiowas, Kickapoos, Osages and in fact all blanket Indians practice this ceremony. Sometimes at these cries the squaws will cut themselves on the arm, neck or face and smear the blood over themselves, to look as hideous as possible. I have been told when the Indians punished an enemy that was a captive, the squaws were more barbarous than the bucks. When an Indian died, his horse and his dog were killed and with his gun all buried with him. This custom was practiced among the Kiowas and Comanches and likely all the wild tribes.

The Indians in west Texas used to throw up their smoke signals by circling a hide or blanket around a fire built for the purpose. The smoke would shoot up high in the air for a second. One or more smokes was used for different signals, Straight smoke, only one, was to let each other know where they were; another smoke, danger or warning, or to come together as they understood the signal. I have seen these signals for twenty miles.

In 1869 I hired to a man by the name of Keener to go to Colorado with cattle. Keener lived below the Littlefield bend on the Brazos river. The next spring two other cowboys and myself returned home. After crossing the plains we camped at the head of Concho river and remained there two days. While there we concluded to look after some wild hogs as we wanted some fresh pork and it was about one hundred miles to a settlement. We saw some signs of hogs, with no owners in sight. Tom McCoy and I went to hunt for the hogs. Tom went down one side of the river and I the other. Soon Tom found a bunch of hogs, all blue, and all sizes. He shot and wounded one and every hog took after him and he climbed a tree. He killed several but they stayed with him. The more he hallooed and yelped, the more they rallied, and more hogs came. I ran to him to help skin the hog, but when I got in sight they came for me and I had to go for a tree like a scared cat. We hallooed and yelped and shot but they stayed with us. After a long time they started away, and to give them a good scare, I gave a loud yelp but they came back and stayed with us until sundown, when they went away; but we kept quiet this time, so we got away all right. Our comrades in camp told us they were musk hogs. That done us in the hog business.

We were advised at Fort Concho to lay over a few days, that the Indians were raiding the settlement and we might have trouble before we got through. After two days we went on. The first thing we saw when we reached the settlement was a bunch of Indians all around a house yelling like demons. . They had killed a calf near the house and ate it raw. The woman barred up the door, got all her children in the house but one girl, who was away from the house. She hid when she saw the Indians coming. As soon as we saw the situation we concluded to make a break for the house, and help protect the people. We tied our lead horses in the creek and got as near as possible and made the run for the house. When they saw us coming they retreated, and I left the other boys and followed them a short distance. I ran to the house and hallooed, and an old gray haired lady opened the door and came out. She gave me an old-fashioned hug like my grandmother used to do when I was a little boy. There were children from fifteen years old down, came out, and the girl who was hid out came in, then the old lady and the married daughter went to getting dinner. The girl went with me back for the horses we hid. We stayed until next morning and left for home.

When we got back near Dublin I saw a sight I shall always remember. We stopped at a house to get dinner and there I saw three sisters with seven little babies, all from one month to seven weeks old.

On my trip to Colorado I owned *and* rode the horse that Pleas Boyd rode when killed. He was afraid of Indians and proved very valuable in a good run or at least I found it so at one time. He was killed by the Indians on Kickapoo creek in Parker county after I got back home. Pleas Boyd had named him Black Hawk, and on one occasion he carried me from almost among the Indians in good style. When we started to Colorado, Keener had bought several hundred head of cattle in Jones county, I think, and sent two other boys and myself to receive them. When we got near our journey's end, we came up face to face with a bunch of Indians. My lead horse was the pack horse. The other boys had left their lead horses with the herd, about a hundred miles behind. All our bedding, our cooking utensils and grub was on this horse when we saw the Indians. They made a charge on us and we had to run. It was our only show. By the time I turned the pack horse loose and got started, the Indians were nearly on me but Black Hawk was scared and made a good run. If I had been on a different horse they would have gotten me, but Black Hawk was quick and afraid of Indians. In about half a mile I caught the other two boys, and here came the pack horse. The pack had turned and everything was gone except the sack of flour and it was going too. In about two miles the Indians stopped and we slowed up, the pack horse still going at full speed. After the Indians left us the two boys went back to look for, and gather up the pieces, while I followed the pack horse. He ran back to where we stayed all night. He was stripped of everything but the dough. The flour and sweat had formed dough all over him, I had a time washing it off. After a while the boys come, they got nearly everything; we stayed all night and went the rest of the way all right.

I will relate a circumstance that happened in 1861, it has been forty-eight years ago and I have forgotten names and the places but I remember the circumstances and give it as I recollect it. I think it was in Bell county or in that vicinity. The Indians made a raid, killed several people and carried off two girls. There was a man called the still-hunter, that hunted Indians by himself most of the time. He had a shot gun made to order. He scarcely ever rode horseback, and could walk thirty-five miles a day on an Indian trail. A party of men had followed the Indians and lost the trail but the man on foot kept the trail, he was well experienced and knew just how to proceed. Indians, after they make a long run and think they are out of danger will stop several days to rest. The man followed the trail several days and on the fourth day he saw a smoke several miles away. The Indians had killed a buffalo and were roasting it. Late in the evening he located their camp and laid his plans for the night. After dark he went up near the fire. The Indians were feasting and having a good time. The two girls were tied and deprived of all their clothing, even stripped them of their shoes. After he had crawled within a few yards of them he fired both barrels of his gun, killed two Indians and the rest all ran off. He ran to the girls, cut them loose and told them to run after him, but they were hardly able to walk and couldn't run, he dragged them away from the fire into the dark and by hard work he got them about three miles away that night, and hid them in some sage brush and stood guard all next day. The girls were blistered from head to foot and sick. He gave one of them his shirt and the other his over-pants. Late in the evening he got them to a small creek. He had them bathe and he caught some frogs and craw fish and fed them. That night he got them two or three miles further. The country was full of thorns, cactus and flint rocks, and made it difficult for them to get along. He was afraid to shoot for game, so he fed them on anything he could get. The first week he got them about twenty miles. He then killed an antelope, took the hide and made the girls some moccasins, that is he whanged up something for their feet, and made a jacket for the girl that had the pants. He then killed a deer and made himself a jacket out of the hide. They were all now well dressed, had plenty of meat that he could cook. They traveled in the cool of the day and some at night. After three weeks watching and taking care of the girls like a mother would a child, he got them to the first house or ranch where he had them taken care of until he could get them home. The readers of this tale of the still hunter might think that man one in a thousand, that would do as he did, but I will say to you that it was in the heart of every old pioneer of Texas to perform just such deeds as this when it was necessary to protect the women and children from the red demons. Fear was never considered.

On the Colorado river the Indians captured a woman, just after the war. A cowboy saw them, ran to them and killed the Indian the woman was riding behind, and hollered to the woman to run after him. She obeyed and got into a thicket. The cowboy stood them off and took the woman safely home to her husband and children. Some reader may say he is one in a thousand, but I will say that if all the noble deeds were recorded that were done by the brave boys of Texas, you would be astonished beyond all reason. All the deeds of the brave boys never can or ever will be written. God bless their brave souls. When I hear one say they were a set of outlaws, they hear from me in quick style. No old pioneer boy will listen to any such talk.

I saw a man in Sansaby county that was acquainted with the cowboy that rescued the woman from the Indians. He said he was as fine a young man as there was in Texas, and that he commanded a squad of men in an Indian fight, which was the hardest fight ever fought on the Colorado river.

As for the mothers and young women of that day and time, God bless them, more noble and pure hearted women never have existed and in their way of doing things were just as heroic as the men. While their acts and deeds have never been written, the experience of the women writers of this book is the experience of thousands of others and will go down for generations to come.

I think it was in 1867 a great many took the gold fever. It was reported that there was gold on Peas river. E. C. Hart, J. P. Hart, uncles of mine, Jim Thompson, Bullion Shields, myself and several others, eleven in all, two negroes in the crowd, one of the negroes was hanged later for murdering a man south of Weatherford by the name of Bird, the next year we started, I think, in May, but am not sure now, to the Peas river to hunt gold. When we got to Red river it was bank full; it had been raining for several days and we could not cross. As the Peas river was on the north side of Red river we concluded to wait until the river went down to cross over into the gold field. Jim Thompson, Al Coffman and I swam the river as we had nothing else to do in particular. One of the negroes stayed where we left our clothes and fished. We had gotten over the river and was resting. The negro hollered at us to come back over in a hurry as we were needed. So we swam back and he told us that he saw three Indians ride into a thicket about a hundred yards from where we were resting. The reason he did not tell us that Indians were near us was he did not want to excite us, as it was dangerous swimming. We went to camp and Bob Hart and Al Coffman went to a mountain about a mile away to investigate. They came back and reported that there was several thousand Indians three or four miles back. We concluded to make a run which was our only show, as we knew we were discovered, and the Indians would swim the river at some point that evening or night. We started about two o'clock and ran all night. There were no roads in that country then and we had to do the best we could. When we came to a creek or canyon every man was at work, no time to lose. We ran until about four o'clock next morning. Our horses had given out and we turned one mile due east off of our course and stopped near a creek and some timber about eight miles from Jacksborough, the place we were making for, but we could not go any farther. Just after we stopped we heard the Indians pass us. But our turning east saved us. The Indians went on to Jacksborough and attacked the post about daylight. There were about eight hundred of them. The soldiers and citizens were taken by surprise but the soldiers got out their cannons and soon backed them off to the opposite side of the town from the soldiers. The citizens had a considerable fight. The women and children were placed in the center and protected until the soldiers got in shape to assist them. We heard the cannons next morning and knew the Indians were between us and Jacksborough. We ate a hasty breakfast and turned our course for Decatur, but only went fifteen miles as our stock had all given out.

We saw several men in the evening that were in the fight. Two of them had their faces powder burned. We learned after the Indians had surrendered, that a chief told the captain that they were aiming to surprise the soldiers in a few days with two thousand warriors, and they tried to catch us to keep us from giving the alarm. There was one company of soldiers and about one hundred citizens at the fort. If the Indians had carried out their plot, all the soldiers and people would have been massacred. Our gold hunt saved them. If we had crossed Red river and drove in among them I guess we would have gone up the spout. If they had caught us the night of our run, I suppose we would have gone up salt creek or had they known where we were next morning we would have gone down the valley or some other country.

That fall, in September, we organized a company to go to the Double mountain fork on the Brazos river to hunt for gold. Col. Shields was our captain. We had eighteen men in our company. We had pitched camp and all scattered out prospecting. Three others and I became thirsty, we went to a canyon and drank out of a buffalo track. One of the boys discovered a white rag hung up on a small elm tree. At first we concluded to go and see what it meant, but took the second thought and did not go for fear it was a trap.

Next day we went to investigate and the rag was gone. A half circle of a canyon was around the tree, in the canyon were lots of Indian tracks, and down the canyon were lots of horses tracks, and then we saw what we would have gotten into. There is no doubt but what we would have all been killed if we had gone on to where the rag was. The Indians were secreted in a position where we would not have discovered them until too late, and then the writer would not be here to tell the tale. We did not get as far as the Double mountains, on account of two of our horses getting rattlesnake bitten. We were on the Big and Little Wichitas mostly. We found some copper ore and other signs favorable to us. On our way back to Jacksborough, two freighters drove up with three wagons and teams. The owner of the extra team lay on his wagon, dead. In the night one of the men woke up with a pole cat on him. The second man discovered what was up and was assisting him to help get loose from the cat when the third man woke up, thinking the second man was an Indian, grabbed his gun and shot him. The cat had bitten the man on the hand and face.

The next year we organized a party of forty-two men to go up on the head of the Brazos river in search of gold and when it came to taking a trip, gold was there. So far as I was concerned, gold or no gold, the trip was my greatest pleasure, for I was always ready to court danger on a trip of this kind. Nothing very unusual happened until we got near the mountains where we expected to go. Some of the boys reported seeing Indians, so we were all ordered to be on the lookout.

We expected trouble in some shape from them. Several days and nights passed but all was quiet and we began to think the Indians had left the neighborhood. One evening we camped about a half mile from the river; there was an old buffalo trail that had washed out and made a ditch about six or eight feet across and about six or seven feet deep on an average. The ditch was about a mile long and had some mud and water in it. The grass had grown so tall that in places it lapped over the ditch. We camped on the north side of the ditch, so as to have one side for protection, our wagons in a circle on about five acres of ground and our horses in the circle all staked out. My wagon, or the one I was with, stopped near the ditch. There were three wagons on the side next to the ditch so this only left three sides to guard; we had four men on guard, one on each side, and one general roustabout, as he was called. Gabe Hart and myself made our bed between the wagon and the ditch; Jim Hart and several others had beds near the ditch, all within ten or twelve feet of each other. Three Indians came up the ditch to where we slept, crawled out of the ditch between our beds and the horses. My horse, Black Hawk, that I have spoken of before, nearly broke his neck trying to get loose; I ran to the frightened animal, got him and brought him to the wagon and hitched him there. Everything was confusion. The guards awoke all the other men so that the number of guards might be increased. The Indians were inside of the circle formed by the guards, lying among the horses; not thinking that the Indians were so near we looked in other directions for them. The horses all became quiet except Black Hawk, which by this time had grown quieter. In about an hour we all laid down and were soon asleep, except the guards.

The next morning three of the best horses had been cut loose and were gone. No one knew just what time or how they were taken. George and Henry Featherkiles and Spears owned the missing horses. It was a certain fact that the three Indians by their craftiness outwitted forty-two men, while eight men were on guard.

On this trip we surveyed several valleys of land, built many air castles, we returned home aiming to come back soon, but this was the last time I ever made an expedition of this kind.

Just after the war I went horse hunting down in Hood county somewhere between the Comanche Peak and where Bluff Dale now stands, with a man named Rufe Green. We stopped to gather some black haws, in roaming around we separated for a while. I was passing near a steep bluff on the sides of which were some large shelving rocks; the dirt had fallen from between them forming a complete shelf. A big, long horn steer took after me and I ran to the shelf and climbed between two of them, the space was so small I had to lie flat down. The space between the rock was just large enough to admit my body. I crawled in with my face next to the wall and could not turn over without rolling out.

The brute could not reach me with his horns, but he got very friendly and began smelling and then he licked the clothes off my back. I kicked, hallooed and squirmed but the mad beast hung on just the same. Green heard my cries and came to my relief. He got on his horse and ran the steer off. When I got out I felt foolish.

I learned some two years ago that Rufe Green is still living near Weatherford.

One time when I was on Buck creek in Erath county, I was helping a young man whom they called Red Shirt, break a young horse, when we got out of the brakes the horse balked while we were trying to get started to run from a bunch of Indians that were about to overtake us I got behind the horse and whipped and slapped him but it did no good, he stood as still as a stone, the Indians were getting too near to suit me, and being determined to stay with him, Red Shirt pulled out his six-shooter and hit him over the head, when he lit out. I never saw a horse run so in my life, I was left behind, I kicked, spurred, whipped him with my quirt, but it seemed I was running a slow race. I leaned over forward that I might be as far from the Indians as possible. I could almost feel the arrows sticking in my back. I jumped forward and leaned away over. We ran down the valley and headed for a house, and the Indians quit us. This was a race for life but we came out unhurt.

In winter most of the people hunted the buffalo in order to get a supply of their meat, tallow and hides. They would take from two to four yoke of oxen to a wagon and bring all one or two families needed; fat cows made the choice beef, only the hind quarters were used and the tallow was saved to take back home. For camp use, ribs and hump were the choice. The hump was streaked with fat and when hung on a stick before the fire and broiled, made the finest eating in the world. Buffalo meat never made any one sick, a person could eat all they wanted at any time. It almost makes me wish for the good old times back again, when I think of the good times I had on the buffalo range. Only the hams of the cow were used, the fore quarters were left, and two or three hides to the man were brought home.

A buffalo hide would make a fine bed. I have slept many a night on a bed made of a buffalo skin. They made fine beds for children to sleep on.

The hams were salted down whole, and left for a week or more and then scalded in a pot of hot brine, then they were resalted, and by the time they got home the meat was in fine condition. The hams could be hung up and the meat was fresh and sweet all the next summer.

The color of a buffalo is a dark brown, there are bones in the hump like ribs only they are straight, growing from the point of the shoulders, back as far as the short ribs and forward on the neck, longer at the shoulders and tapering in length each way. A bunch of wool grows in their forehead several inches long and after they are over two years old, a ball from an ordinary rifle wouldn't penetrate it. To shoot one through the heart you had to shoot just behind the point of the shoulder and low. If you were to shoot one as high as you would a cow you would shoot it through the lungs. I have seen them shot through the heart, a death shot, and not even flinch, but feed ten or fifteen seconds, lay down just as though nothing had happened and be dead when they hit the ground; they cannot be killed as long as they are mad. If you shoot them a dead shot they never die until they cool off. In feeding they grunt like a hog and are as bad to wallow as a mule, and like wild geese they travel north in the summer and south in the winter. They are always a few scattering bulls in front of the main herd, from twenty to thirty miles. One bull was never whipped by another but always fought to his death. Sometimes there would be as many as two hundred thousand in a single herd. The bulls were generally on the outside of the herd to protect the cows. We always had to go on the opposite side of the herd from the wind to get within shooting distance and had to crawl on the ground sometimes from a quarter to a half mile, according to the lay of the land. This was when the cap and ball rifles were used, and we had to get as near as eighty or a hundred yards to them to get a shot.

The first time I went buffalo hunting I was "Pocket Change" for all the crowd, all old timers know what that means. In any crowd there was always one for "Pocket Change"; among the men, women or boys some friend was singled out as "Pocket Change." I was opposed to drinking, gambling and swearing and I never had a case in court in my life, yet I was sometimes accused of being as bad as Peck's Bad Boy, and in my bashful teens men and women both made "Pocket Change" of me.

I had been on the range a week and all had killed a buffalo but me, and I was teased nearly to death, but being determined to kill a buffalo, although I knew when I did they would say I found a dead one or something of the kind, and all kinds of tales would be told on me about the buffalo I had killed.

One day my uncle Bob Hart and Jim Esque and myself went out and found several bunches of buffaloes, my uncle picked out a bunch for me to go to the easiest place to crawl to, and then told me how to go and they left for another bunch. I crawled for about a quarter of a mile, flat down like a lizzard, and got nearly close enough to shoot when a big black pole cat came along and stopped right in front of me and would not go any farther. I laid still a long time thinking he would move on but he came a step or two nearer me, then he would pat his feet and dance, come again and pat. He got very friendly and it seemed he wanted some fun, and I began to think he was going to have it. I tried several things to make him change his course, but he got bolder and would pat his foot and look saucy, and the thing was having all the fun, and the fun was becoming unpleasant for me so I pulled away and shot my tormentor to pieces, and away went the buffalo, all the buffaloes in the neighborhood got a move on them. I backed up against a mesquite tree, stood there for about half an hour and felt like a fool. If I had had any religion I would have lost it in a wad right there. Of course I was accused of not knowing a pole cat from a buffalo. They told that I cut the hams from the pole cat and hung them on a tree just like they did the hams of a buffalo. They said I stretched the hide to take home with me for a keepsake to show the first buffalo I ever killed. They told it on me that I crawled nearly a mile to a bunch of buffalo and lost my gun and when I got to the buffalo, I raised up, took sight a long time and then said, "Gee whiz, I have lost my gun."

The next morning uncle said I should go with him and he would assist me in getting a shot. We soon found a fine bunch but had to crawl a long way, flat down. Out there in the mesquite country there is a cactus that grows nearly flat on the ground, perfectly round like a plate and full of big thorns. I came to a place where I could crawl on all-fours for a short distance, then had to lay down flat again, so I laid down flat on one of them. It was as big around as a large saucer. Now, just imagine what I did without me telling it. Every buffalo for two miles around was on the run in less than five minutes. I dreaded to go to camp that night, but I went in and as usual, took my medicine. The next day it was agreed to move camp further up the Wichita river, as I had scared the buffalo off, and all would hunt until noon and I was to get dinner. I had the bread on, and a hot fire on the lid and a big pan full of meat frying, when I saw a bunch of buffalo about a quarter of a mile off, and easy to get to. I grabbed my gun and went. When I got close enough I fired. I broke the leg of one and killed a calf about two weeks old. This was worse than ever and I thought I would say nothing about it, but two of the boys were crawling up on them when I killed the calf. We went to camp and the bread and meat was burned to a coal, and I told the boys they could get dinner or do without. The calf, to hear them tell it, weighed fifty tons. The next day I killed two buffalo, one antelope and one turkey. From then on I was about a third class hunter. One Day Bud Eddleman and Al Coffman were crawling up on a buffalo herd. Eddleman had the advantage of a dead buffalo and Coffman a bunch of prickly pears. When Eddleman was about fifteen steps from the dead buffalo, Coffman hallooed to him to look out. A wolf on the opposite side of the dead buffalo made at Eddleman and got in about four feet of him. Quick as lightning he drew his six shooter and shot it in the mouth. This was all that saved him. The wolf was a large black one. Buffalo hunting was very exciting and sometimes dangerous. One morning just before daylight I went for a bucket of water. As I got back up the bank of the creek, I met Eddleman; he did not know anyone had left camp. Thinking I was an Indian he drew his revolver to shoot just as I dropped a tin cup. He heard the cup fall and that is all that saved me.

PIONEER DAYS IN THE SOUTHWEST

At one time I got awfully scared. I had wounded a buffalo that ran up a ravine in the direction of a cedar mountain. It went up on the side of the mountain, but I did not know it had left the ravine. Looking every step to see the buffalo, I got within about forty yards of where it got up, ran over old Cedar Top and started a boulder down the mountain toward me. I thought it was after me and I wheeled and ran for dear life, about sixty yards and climbed a mesquite tree as fast as I possibly could. The boulder which I thought was the buffalo, only rolled as far as the foot of the mountain, and I had all my fun and climbing for nothing. I got down and looked around to see if any of the boys were in sight.

One day when I went out it was cloudy and I soon got completely lost and I traveled all day. All the mesquite flats looked just alike and I had nothing to guide me. Just at night I saw two of the boys going to camp. I fell in with them but did not tell them I was lost.

At another time, another man, I have forgotten all his name but Ed, he was afterwards killed by Indians, he and I went out one morning, it was snowing lightly. We wounded a buffalo and it went over the mountain. I went around one side and he on the other, but we missed each other and I got lost and traveled all day. The snow was about an inch deep. Darkness came on and I went into a small canyon, gathered up a lot of dead mesquite wood and got a rat's nest out of a hollow chunk to build a fire. I had passed a small mountain about a half mile back. Buffalo hunters when anyone failed to come in, always went to the highest place and built fire for a guide. I concluded to go back to the mountain and see if I could see any fire signals. I looked back towards my fire and just below it about a quarter of a mile I discovered another fire. I made up my mind at once I had discovered an Indian campfire, and I would go as near as I could and see what I could find out, I went within a few yards but could see no one. I concluded the Indians had discovered my fire and were watching it. I at last ventured up to the Indian's fire and looked at the tracks; they looked like shoe tracks, Indians wore moccasins. I didn't stay long but went back to my fire to see what I might find out. After I froze out I went to my fire, punched up the chunks and warmed myself. I was there probably half an hour when some one hallooed, "Hello John!" It was Ed, he had gotten lost too, and did the same thing

I had, even gone to the mountain to look for signals. He came down to the fire where I was and we shook hands and were very glad to see one another. He had killed a rabbit and saved it, for he thought he might need it. We broiled the rabbit and had a feast. It cleared off and the sun came out bright and nice. We went to a mountain and could see the timber on the Wichita and by noon we were at camp.

Four of us went out and killed eight buffalo all near each other. By the time we got them dressed it was dark and we concluded to stay all night as it was six or eight miles to camp. We broiled the hump for supper and breakfast. For a bed we laid a hide down hair side up and covered with a hide hair side down. It was cold but we kept warm and enjoyed ourselves fine. This was rather an uncommon bed and I don't expect a New York dude would care to sleep under a bed blanket made of green buffalo hides.

In 1874 I went on my last buffalo hunt. This time we hunted on the Clear Fork of the Brazos river about forty or fifty miles above Phantom Hill, an old fort. The hunt was about like all buffalo hunts—very exciting and plenty of enjoyment.

In 1876 about all the buffalo were killed. Hunters killed them for their hides and tallow. As soon as the buffalo were all gone the Indians were conquered. It was their bread and meat.

Three years ago at Clarendon, Texas, I attended a reunion of old Parker county school mates. I met old boys and girls that I had not seen for thirty-five or thirty-eight years. Everybody looked strange. Many of them were fathers and mothers and some were grandparents. W. J. Parsons and L. C. Beverly arranged for a regular picnic. There were about 200 people at the picnic. We played marbles for four days straight, just as we did when we were boys. We went to Charley Goodnight's buffalo ranch. He and his wife were both old timers; they furnished us with hacks and teams to drive around. There were about sixty buffalo on the ranch. It made me feel like old times when I met the old timers and the buffalo.

Antelope hunting was one of my chief sports, but it took experience to kill them; their feeding grounds were always on the high divide or level plains. They would discover an enemy as far as they could see. I have seen from one to several hundred in a bunch and in a day's travel could see thousands of them, but they are all scattered now, only a few in number, and will soon all be numbered with the buffalo. It was nearly impossible to get near enough to get a shot at them on high ground. I have put a red handkerchief on a gun rod stuck it in the ground, lay down in the high grass and they will come for more than a mile on a run within sixty or eighty yards of the signal, and make a complete circle around it and make a thorough investigation. I have laid down in the high grass, kicked up my heels occasionally and caused them to come. They make a half circle, and then stop for a few seconds and then full circle and repeat several times before they leave. Near Pilot Point, when that was a frontier country, there was a woman who enjoyed antelope hunting. I have heard that she killed five in one day, besides at other times she killed a few deer and some wolves. A woman in Young county killed a panther with a smoothing iron. She was ironing after wash day. She had laid her baby, about two months old, on a pallet and when she looked around a panther had grabbed it. She got in a lucky lick on its head and pounded it until it was dead. The Indians attacked a house and the woman killed two Indians. The people after this gave her two fine six shooters. After this I learned she lost her mind and was sent to the asylum.

While I have not written all my frontier experience, and some that was very dangerous and some that seems unreasonable, yet if the danger could be taken from the women and children, and replace the buffalo and antelope, the Indians in his war paint, I would love to have the pleasure of experiencing the old times over again. While I have run for dear life, and been almost scared to death, even when I stood up before the preacher and he commanded me to take her by the right hand and scared until my breath was almost gone, I would like to experience the old times over again. But the beautiful frontier that was at one time the most beautiful park on earth with its great herds of cattle that roamed over the beautiful prairie, the buffalo of the plains, the antelope on the high ground, the Indian with his spear, bow and shield and the old pioneer people of that day are nearly all gone. When I look back to the days of my boyhood, with some of the dear boys who yet live, it seems life was only a dream. I am now fifty-eight years old and I and the pioneer boys and girls who yet live will soon be no more in this world; but I hope when we pass over the river that we can all say that the acts of our lives have been more than a dream.

PIONEER DAYS IN THE SOUTHWEST

CHAPTER VIII. BY JAMES D. NEWBERRY, CHILLICOTHE, TEXAS.

My father, R. C. Newberry, and two brothers, Campbell and Ross, moved from Tennessee to Parker county in the fall of 1859, stopped and on the Brazos river, on what is now known as Big Valley near the mouth of Kickapoo creek. My father lived there until 1861, then bought land on Grindstone creek west of the town of Weatherford, the county seat of Parker county. My uncles, Campbell and Ross, bought on Kickapoo creek near where Lapan now is. It was then Erath county, since then Hood county being formed, their old place is now in Hood county.

In 1861, in December, I think it was, the Indians killed a man by the name of Brown, fifteen miles northwest of Weatherford. The same day they killed a Mrs. Sherman about six miles from where they killed Mr. Brown and drove off quite a lot of horses. The men of Parker and Palo Pinto counties came together and followed the Indians. They were joined at or near old Fort Belknap by Captain Saul Ross with his Texas Rangers. They followed them up between Peas and Red rivers. There they overtook the Indians and had a fight with them, killed a number and captured Cinthy Ann Parker from them. She was captured when small in Anderson county, she was then grown and the wife of the Indian chief. Quanah Parker, the Comanche chief, is her son.

In January or February they killed Bill Youngblood in Parker county, about twelve miles northwest of Weatherford, and scalped him. The neighborhood gathered up and followed them out to Keechi valley, there they were joined by the men of that neighborhood and overtook the Indians, killed two of them and got the scalp of Youngblood and got back in time to put the scalp on his head before he was buried. My brother, then a boy, saw the Indians that morning before they killed Youngblood. My brother's name was C. H. Newberry, we called him Huse.

In February I joined Captain John Tubb's company and on the first of March started off to war. That was March, 1862. I was not eighteen years old then, but while I was in the army, which was until June 1, 1865, the Indians killed a number of men and women in Parker and Palo Pinto counties, and stole lots of horses and cattle of the white people. A number of the whites went east to get out of danger. The Indians used bows and arrows and spears as their weapons of war The arrows were made of dogwood switches with an iron spear at one end and feathers at the other. The spear was from four to six feet long with an iron spear at one end about ten inches long. They could shoot those arrows at sixty yards with correct effect and keep from two to three in the air all the time.

During the war the old men and boys formed companies and scouted on the frontier. My father and brother, Sam, belonged to the scouts. After I came home from the army, I had no horse, for the Indians had stolen mine. I went to work and bought a horse to ride, and we used oxen to work, both to the wagon and the plow. I broke some land and sowed wheat which yielded me a good crop.

In the fall of 1865 the Indians made a raid and came through our neighborhood. G. W. Light, H. B. Moss, C. E. Rivers and A. J. Gorman got together and followed them five or six miles and rode right up on them. The Indians made fight and the whites fell back for a better position. They did not find it and the Indians killed Gorman and scalped him, but didn't get his horse. He was a widower with four children, his wife had been dead three or four years, so the little ones were left on charity, but they were cared for. We followed the Indians but failed to overtake them and gave up the job.

Young men and young ladies of the frontier would ride horseback to church and go visiting twenty-five miles. The men with a family would go in ox carts with the women and children, but they had their guns and pistols along. The preaching would be under a brush arbor, and all did enjoy each other's company. All were on an equality and if a man got in a tight place they came to his help with labor and money. If we wanted money we got it without note or security. Clothing was homemade for both men and women. The ladies would go to church dressed in these homemade dresses, the men wore the homemade suits throughout. We didn't need much money those days.

In December, Isaac, Betty and I got out as many grindstones as three yoke of oxen could pull and-went down east and sold and bartered them out and brought back a load of bois'd'arc wagon timber. On December 28, 1905, I married Miss M. E. Porter and moved to ourselves, and went to our work in January, 1866. The Indians came in the neighborhood and stole a lot of horses but did not kill anybody until about the first of March. They then made a raid down southeast of us and stole a lot of horses on Sanches creek. They killed a Mr. Savage while he was in the field plowing. Some of the horses they were driving ran into Mr. Savage's yard as the yard gate was open and Mrs. Savage drove them off with a gun. The horses belonged to Mr. John Hart. Then the Indians went over on Patrick's creek and killed a brother to the Mr. Savage they had killed. They also got some tools and three of Mr. Savage's children and carried them off with quite a lot of horses. My mother saw them as they passed my father's house and counted them; there were twenty. The Indians passed out to Rock Creek. There Mr. Fuller Millsap lived and had his horses in the field. He saw the Indians trying to get them and he and a negro woman took their shot guns and ran them away from his horse. So a party of men followed the Indians out to Keechi creek, but night came on them and forced them to abandon the trail. It was a hard task to follow their trail as they would go over the roughest places if they knew you were following them and when in the glades or gravelly places they would all scatter and make it almost impossible to follow them. We made good crops of wheat and corn, as that was all we raised then, and only a small patch of each; all was a good price. People depended on cattle mostly, the country was full of cattle for the range was fine then. The Indians did not kill any other people in the year 1866 that I remember of now, but they stole lots of horses and cattle, and kept everybody on the lookout for them all the time.

In the year 1867 we had some more trouble with the Indians. It was in May, I think, when the Indians made a raid, in daylight, on the Brazos river in what is known as Littlefield Bend and stole a lot of horses. The men and boys started after the Indians and run them five or six miles. They had a running fight, as the Indians kept going. Some of the men crowded them and had a hard fight. Of course some of the horses of the whites were speedier than others. The men shot two or three Indians off of their horses and went on in pursuit of the rest, thinking they had killed those that fell off. The white men behind seeing them as they passed also supposed them dead, and went ahead for they heard their comrades shooting and hollering. The front one ran upon the Indians so closely that the men shot two and the remaining four turned on the whites and Mr. Monroe Littlefleld was in front of the whites so the Indians directed their fire on him. One of them had a six shooter and Mr. Littlefield said he saw the Indians would shoot him and he had emptied his pistol, so he whirled his horse and threw himself forward on its neck. The Indian shot him at the point of the shoulder blade, as he was leaning forward, the bullet ranged upward and was taken out just above the collar bone on the same side. This ended the fight as they had to care for the wounded man. They carried him to Mr. Fuller Millsap's house and sent for a doctor who cut the ball out.

When they went back to scalp the Indians, as was the custom, one had made his escape, so they only scalped two. I saw them myself, they got a lot of bows and arrows, shields and spears, Indian saddle blankets, lariats, bridles and one bridle had $15 worth of silver plates on it. They sold all the things at Weatherford to pay Mr. Littlefield's doctor bill. The doctor paid $20.00 for that fine bridle but he could afford it as he was the doctor and Mr. Littlefield was protecting the doctor as well as everybody else. It took Mr. Littlefleld a good while to get up, he never did get entirely over it. The neighbors worked his crop and Mr. Millsap would not think of taking pay for what time he was there. Though wounded, Mr. Littlefield was one of God's noblemen; he has long ago crossed over and answered the last roll call. He was an ex-Confederate soldier.

There were frequent raids made by the Indians in the county, but can't think now of any particular one, they were so common; they generally came in during the light of the moon, and were as sly as a wolf. Most everybody went well armed for protection. People would go to church as much as they could, but the preachers were shy of coming to preach. The Methodists would preach once a month, generally at my father's house, though he was a Cumberland Presbyterian. He and my uncle sold out in Erath county and moved to Parker on Grindstone creek, near father's place and bought land. My father bought two places and made partial payments but never did give his note in either trade. That showed the confidence people had in each other. The old settlers of the frontier have nearly all crossed the River of Death. The country changed from a range and cattle country to one of farms. We thought at the first settling of Parker we

4

never would want much land, as the country would always be open and free range, but alas! we were doomed to a sad mistake. It does seem that the old pioneers should be honored, but people of today seem to think they did nothing but their duty, and don't respect them as they should for risking their lives and the lives of their families for the protection of the country.

1868 was a dry year. We had no rain from August 1867 until in March 1868. No wheat nor much oats but some very good, corn was raised. In April the Indians came and stole a lot of horses—got all my brother-in-law had. We followed them all day and went to a certain Crossing on Keechi creek that they often went through, thinking we would head them off. Two or three other parties got with us but when we got there they had passed on. It was now after dark but in the full moon. We nor our horses had had anything to eat since early that morning, we went about three miles, hobbling our horses out and took turn about watching them. Next morning we started home and never got anything for ourselves until we traveled ten miles and through a heavy rain. I never got home till dark, but we failed to come up with them, and in May they came again and stole a lot of horses from some cow men. My father lost a fine mare.

On Saturday, July 4, there was preaching at my father's. Most of the neighborhood was out. All was peace and quietude. Wife and I went to her sister's, Mrs. H. E. Moss, and in the evening Mr. Moss and I went up to my home to see about my things, having decided to stay with them that night. We had just got back to his home when we heard shooting commence up the creek and we ran our horses as hard as they could go. It was about two miles to where our brother-in-law Light lived. It was about sundown when I rode up to the gate and hallooed. He answered and we turned and rode up to where he was. He was shot, his wife and one child dead and another wounded. He lived about three hours. That was a trying time. The Indians went south about four miles to Mr. Eli Cox's on the big road, and got over into his cane patch and was eating it when Bob Tinson came from Weatherford. It was now dark, and they shot at him. He ran into the woods and got away from them without being hurt.

In about two weeks after they had killed the Light family, his father, G. W. Light, went over for his son's chickens and put them in a two horse wagon at night. He went home with them and took a young man named Spurgeon. The Indians ran upon them just after they had crossed Grindstone creek and got their wagon and horses. Of course they did not take the wagon but cut the horses loose, and took them and the one Spurgeon was riding. He left his horse and got in the bed of the creek. Mr. Light shot one of the Indians but they carried him off. Mr. Light and Spurgeon went to the home of old Judge E. S. Porter and escaped from the Indians.

In June of 1868, Mr. W. H. (or Tip) See. he went by the name of Tip, lived on Buck creek in Palo Pinto county, was out hunting stock. The Indians killed him and his body was not found for three days. He was a good man, and we were in the same company in the army. He was one of God's noblemen, and all the company loved Tip See.

All the neighborhood left the settlement then, but my father and old uncle Jesse Perkins, and myself, stayed with what little we had, but we had a lonesome time. In the fall my uncle, Campbell Newberry, bought a place and moved in about a mile of us and some few came back to their farms.

In 1869 young Elbert Doss lived with his uncle, H. E. Moss, and his brother, John Doss, lived with me. The two Doss boys, my brother, Sam P. Newberry, Milton Ikard, Jr., Tom Cox, Will Gray and a negro boy, Bose Ikard, were out on a cow hunt. I started to them but it rained and I turned back. They went on to a roundup that evening, and struck an Indian trail and followed it about twelve miles up into Palo Pinto county near where Mineral Wells now is and ran on the Indians in a bluff of rock and timber and had a fight with them. They killed young Elbert Doss, shot him with a pistol near the heart and it went clear through his body. The boys put him on the saddle and one rode behind him on the same horse and held him on. They set him up as though he was living, and led the horse that carried him home. They did not get any of the Indians they killed but they saw them fall off their horses. The boys were all young men, none older than twenty years. That was April 24, 1869. In June of the same year the Indians made another raid and stole some horses. They came to my house and my wife woke me up and told me she was satisfied the Indians were around. I got out of bed and woke my brother-in-law, W. B. Porter, we went out and watched our horses awhile and while we were out the dogs barked at something just outside the yard fence by a post oak tree, but we couldn't see and didn't go to it. The next morning there was the Indians' tracks behind the tree. We gathered up some men and boys and started after them before breakfast, as they had stolen several head of horses. There were ten of us and we followed their trail about twenty miles in a roundabout way and over an awfully rough rock country. It rained on us hard and we lost trail of them. While hunting it, seven or eight men came to us. After a long time we struck the trail on top of a high mountain, where Mineral Wells now is. We ran them ten or twelve miles, the woods were miry and boggy. We crowded them so close they dropped all the horses except the ones they rode. In the meantime the Indians passed a house where they had washed and they took all their clothes. We got so close to them that we got shots at them several times, but to no effect. A good many had to stop on account of their horses giving out, so we lost the trail in the cedar brakes of Palo Pinto county. They were a daring, sneaking, nasty foe but we had to contend with them just the same.

I then moved my family to Weatherford, but I stayed out on Grindstone creek, at my father's most of the time. One night brother Sam and I were sleeping out between the houses at father's. Father came out and woke us up saying he had seen twelve Indians pass by, as the moon was bright. He said, "now boys if we watch we will kill some of them for they will come back to the stable for our horses," as we had them locked up in the stable. So we all three went out with our double barrel guns and pistols, and got in the shade of the wheat granary, in plain sight of the stable door and about forty feet away. We waited awhile and brother Sam and I both went to sleep; father woke us up by shaking us, we roused up but in less than thirty minutes were both asleep again; so father woke us up again and gave us a scolding, saying, "go to bed, for you are no good to watch as you both will sleep," but the Indians did not come back; I guess they saw us as we went out of the house.

Along in the fall I moved my family out of town and about three miles up the creek to my father-in-law's; we all lived together for protection.

I had two brothers and a cousin living at my father's.

PIONEER DAYS IN THE SOUTHWEST

In January, 1870, the Indians killed Mr. Rippey and his wife about seven miles north of us on Rock creek. His dogs barked at something about a hundred yards from the house; he went to see what it was as it was across a small field. When he got there the Indians shot and wounded him, his wife seeing the trouble went to his relief, and the Indians killed them both.

In February of 1870 Ed Lamdua lived with Mr. Fuller Millsap and his son-in-law, Joe Loving, and worked for them. One morning he went out to hitch up a team of horses to the wagon, and while he was fastening the breast yoke the Indians shot him, having crawled up close, unseen by him or the rest of the family. Millsap and Loving fought the Indians for about two hours and kept them off. They killed several Indians but poor Lamdun died in about three hours. Millsap and Loving kept them from getting any of their horses. During the fight Mr. Millsap ran out of cartridges for his Winchester rifle and his daughter, Miss Donie Millsap, took some in her apron and carried them to her father; as she was going back to the house the Indians began to shoot at her and as she went through the door of the house they shot an arrow through her apron and it stuck in the door facing.

Mr. Millsap at that time lived sixteen miles west of Weatherford on the Palo Pinto road on the bank of Rock creek, he afterwards moved to Weatherford, Texas, and lived there for about three years, till he died. His widow and two of his boys, Tom and Will, still own the old home place. Mrs. Millsap is getting very old and feeble.

The town of Millsap was called for them; it is on the Texas & Pacific railroad fifteen miles west of Weatherford, Parker county, Texas.

In June of 1870, the Indians killed two young men on Grindstone creek in Parker county. They lived in Palo Pinto county and had started to Weatherford to mill when they were killed and scalped in the big road and their horses taken. From all appearances they fought the red demons a terrible fight. They were not found until the next day. Their names were Cathry and Hale.

All through these troublesome years the people would come together and hold protracted meetings, and camp meetings. They would go in ox wagons and take their guns and six shooters and have good meetings. They sang the old time religious songs. We used brush arbors and split logs for seats. I have seen preachers stand and preach with their pistols belted around them. Everybody was friendly and glad to meet each other.

In the fall of 1870 brother Sam Newberry, W. B. Porter and I were gathering corn in a valley on Sandstone creek with a range of Parasia mountains on the west. We saw fifteen Indians up on the mountain watching us so we stopped and got up in the wagon and sat there and watched them for quite a while, an hour I guess, but they did not come down to us as we were armed and in its field. When they left we went to the house and tried to get a crowd to follow them but they were all gone from home and we three did not feel disposed to follow that number of Indians as they had as good arms then as we did. The Indians went south about six miles and ran two men or a man and a boy, but they happened to be close to a fence and they ran and got into the field but the Indians shot at them several times. The Indians went on south and on Patrick creek ran a boy by the name of Bill Abbott but he got in the creek bottom and hid from them by lying down in a thicket of dogwood bushes; this was late in the evening, next morning they ran a boy by the name Jim Tierrie on the same creek and shot several sheep that Jim was tending, and stole several horses in the country. We followed them but we failed to overtake them, as it was so dry we could not trail them. I used to think I was as good a trailer as ever was.

In the same year the Indians killed old man Leeper northwest of Weatherford and stole lots of horses.

In 1871 they killed a boy on Dry creek fifteen miles northwest of Weatherford, by the name of Cranfill, old man Loop six miles north of Weatherford and a young man by the name of Hemphill fourteen miles northeast of Weatherford. The young man was killed one night as he was riding along the big road, and the same year I think it was they killed a Mr. Blackwell sixteen miles northwest of Weatherford.

In 1871 on the third Sunday in August, Rev. Ben D. Austin held a camp meeting near my father's house, under a brush arbor, and had a good meeting. He was a Cumberland Presbyterian and organized a church under the arbor, with twenty members, and ever since that time they have held a camp meeting in that neighborhood, but only two miles north, they built a church and called it Newberry church. I gave the church six acres of land for a camp meeting ground and cemetery. It is known as the Newberry church. We have a good house and tabernacle on the land now. I live two hundred miles from there now in Hardeman county, Texas, but Parker county, Texas, seems like home to me as I lived in that county forty-seven years. My church membership is still at old Newberry church.

Rev. Ben Austin, A. C. P., and Rev. Jim Hines would hold a camp meeting together every year at the mouth of Patrick creek on the Brazos river, and we would all go down and what a meeting we would have. It was old time religion. We would all come with our guns and pistols and lay them down under the edge of the arbor and tie our horses—or those that had horses—to nearby trees to keep the Indians from getting them. All the old men that attended those meetings have crossed the "River of Death" and we that were young then are now gray and bent with age.

Among those that would camp on the old camp ground on Patrick creek were: Old Uncle Jimmie Strain, Lewis Marshall, Jack Joyce, Grandma Joyce, Aunt Polly Maxwell, Mrs. Hart and her sons and daughters and grandsons. Rev. Ben Austin and Captain Jim Buckner. Those were joyful days, no style nor formality but plain, practical gospel preaching. Rev. Jim Hines was a Methodist and Rev. Austin was a Cumberland Presbyterian, but you could tell no difference but what both belonged to the same church and they did, for they were trying to build up the church of Christ. Those were good times.

In 1872 the Indians did not bother us so bad, but made several raids through the country stealing horses. My brother, Hugh Newberry, lived with me and one day we hobbled our horses near the house; my wife went to the garden for beans, when she came back she told us our horses were snorting and cutting up so we went to see about them and the Indians had stolen my brother's, but did not get mine.

Will now write of 1863 to 1872. In 1864 the Indians killed a man by the name of Berry south of Weatherford on the Brazos river and a commotion came up in Parker county and J. M. Luncky was hung to a post oak limb because the people thought he was trying to betray them into the hands of the Yankees. The negroes caused trouble by some talk and the people hung four negroes to the well on the public square in the town of Weatherford. About that time a strange man came to town and Captain Munroe Upton went to him to find out his business. The man would not give his mission and he and Captain Upton got into a row and the captain killed him. They never did find out who he was nor where he came from. In 1864 the Indians killed two Hamilton boys on Patrick's creek ten miles south of Weatherford and the same day the Indians killed a Mrs. Brown and daughter and wounded another of her daughters. This all occurred on Patrick's creek in Parker county. I don't know Mr. Brown's given name but I was well acquainted with him. We always called him Black Jack Brown; he was in the army at the time his family was murdered. I think the wounded girl died. One of his daughters lives now in Parker county, Mrs. G. N. Pickard, of Spring Creek; I am well acquainted with her.

In February or March, 1864, Uncle Henry Maxwell and his son-in-law, Jack Joyce, being at home from the army on a furlough, went out to see after some stock and ran into a bunch of Indians, and the Indians surrounded them when Joyce was going to shoot at them, and Mr. Maxwell said, "Jack don't shoot; reserve your fire." Mr. Maxwell's horse became scared and ran away with him. The Indians shot Mr. Maxwell in the body with an arrow and Mr. Joyce held the Indians off but Mr. Maxwell died that night. Mr. Maxwell had a son in my company in the army, and one son in the 19th Texas cavalry also. Mr. Joyce belonged to the 69th Texas.

In 1872 I think it was, the Indians made a raid into Hood county and the citizens got after them and crowded them so on the line of Hood and Parker counties, on Robertson creek, they ran them into a bend of the creek and surrounded them and killed them all. There was a squaw with the Indians but the whites killed all not knowing that one was a squaw until after they had killed her. One white man got wounded very badly, I think perhaps he died, I have forgotten his name. I lived about twenty-five miles from the place where they had the fight. Before that in the same neighborhood the Indians killed Pleas Boyd, and at another time killed Nathan Holt. All occurred in the same neighborhood.

In 1867, after the war, the government sent a company of Yankee cavalry to the town of Weatherford, and they took a big hand in punishing men around Weatherford. They arrested a number, I was one but they only kept me a very short time. They arrested Aaron Hart and tied him up by the thumbs and kept him until he fainted. They had quite a number of men arrested for the hanging of Luncky in time of the war. Some men left and stayed until the Yankees were gone. They arrested Mr. Steve Jones and he started with them but did not go far until he ran away from them, he being on a good horse.

We had lots of trouble those days. Quite a number of men were killed in the town of Weatherford. You know every thing was "rolicky" those days. Mat Gibson was running a meat market and a Yankee captain went to his wagon and ordered his men to take a quarter of Mr. Gibson's beef. Mr. Gibson forbid it, but the captain said he would take it, when he went to take the beef out of the wagon, Mr. Gibson jerked out a butcher knife and killed the captain on the spot. Mr. Gibson hid out for quite a while. He had so many friends that he came in, stood trial, and came clear of the law.

I think it was in 1869 when a bunch of cow hunters, twelve in number, in Young county, on Salt creek prairie, had a bunch of Indians run on them and they fought the Indians half of the day. The Indians killed two or three of the white men and wounded all but one of them, when night came on and the one not wounded made his way to the nearest ranch afoot, as the Indians had killed or got the men's horses. It was several miles off. He got help and buried those who were dead, of the whites and took the wounded in, several of whom died. One of the white men now lives near Weatherford, Parker county, Texas. His name is G. W. Lemley.

In 1872 or '73, Captain Warren had a wagon train of twelve wagons, six mules to the wagon and thirteen men as teamsters and a wagon boss. He was hauling supplies to Ft. Griffin for the United States soldiers on Salt creek prairie in Young county, when the Indians, fifty in number, ran on them and killed all the white men, took the mules and robbed the wagons of what they could carry off. They tied the corpses of the white men to the wagons and burned them. The United States soldiers followed the Indians to their camp or reservation and captured the two chiefs, Surtanta and Big Tree and carried them to Jacksborough and tried them for murder. Ex-Governor Lanham was district attorney and Soward was judge. They were convicted and sentenced to be hanged but Governor E. J. Davis commuted the sentence to life imprisonment in the Huntsville penitentiary of Texas. He afterwards pardoned them on promise of their making no more war on the whites of Texas. They kept their promise partly. We had but very little trouble after that. A few raids were made by the Indians after that. When they arrested the Indian chiefs, the white men killed one Indian chief, I have forgotten his name. For this Captain Warren got fifty thousand dollars from the United States government in cash. He being a big man got pay for his property and the poor got nothing.

We Texans had lots of trouble with the red men. There are a number of their raids and killings that I have forgotten, it being so long ago. I lived west of Weatherford on Grindstone creek for forty-seven years, only what time I was in the Confederate army, that being three and a half years. I am to go down in Parker county in August to a camp meeting at the old Newberry camp ground. They have had a camp meeting every year for thirty-seven years. It is in August embracing the third Sunday; has been on that day ever since it began, but most of the old timers have either crossed over the river of death or moved off. Some few remain. Brother S. P. Newberry, Uncle Ross Newberry, John and Will Doss, Bob Strain are about all the old ones living there now, but lots of the children of the old ones. John C. Newberry and Georgie are of the old ones. There is quite a difference now and when they first commenced to hold these meetings. They make brush arbors for camps now as they did on the start, now there is a good deal of style. At first all things were common; when they went they would carry their guns and pistols for protection in case of need. Many at the judgment day will call Newberry camp precious and a glorious place on earth. They now have a nice church house and a big tabernacle on the camp ground. Your humble servant gave the land for the camp ground and grave yard which is a seven-acre site on the Texas & Pacific railroad, it is a precious spot to me. I have not missed a single meeting since the first one, nor won't, I think as long as I can get back there, but I am nearly sixty-four years old, my race is nearly run, I am just waiting for my Master to say, "It is enough, come home to Me."

I hope if this is ever published it will be entertaining to the reader.

PIONEER DAYS IN THE SOUTHWEST

PIONEER DAYS IN THE SOUTHWEST

CHAPTER IX. BY MRS. MARY A. NUNEZ, THORP SPRINGS, TEX.

I was born on Sugar mountain in Lee county, Virginia, in 1848. My grandfather and grandmother on my mother's side were Rev. V. Al and Mary Woodward, or Polly as she was called then. My grandfather was a Methodist preacher. Two of my mother's brothers were preachers also, William and Alexander. My grandmother was a Ewing before she married my grandfather. She also had two brothers, preachers, Jo and Alexander. My father and mother were Isaac W. and Elizabeth N. Cox. My father had one brother, a Methodist preacher. Ivy H. Cox, and two brothers, doctors, James H. and George Cox. They have all passed away long ago. We emigrated to Texas when I was about four years old and settled within four miles of Ruterville, Fayette county, and twelve miles from LaGrange. It was in 1852, as near as I can remember. We came all the way in house wagons. I don't remember how long we were on the road. We stopped at Little Rock, Ark., quite awhile on account of sickness and lost a little sister there.

Our nearest neighbor was a German by the name of Haller. Our next neighbor was John Pain. My mother taught a school at our house and Will and Annetta Huller, and Bob and Mary Pain attended school. Well, when we had been in Texas four or five years my grandfather in Virginia wrote to my mother to pay them a visit and they would give her part of the estate. So she went in the spring. I remember the apple trees were in bloom. She took my baby sister who was nine months old and myself, nine years old with her. Traveling conveyances were not as convenient then as they are now. There was only one railroad in Texas at that time, and that was from Richmond to Houston. My father took us in a wagon to Richmond where we got on a train and went to Houston, and from there on a little boat to Galveston and from there across the Gulf of Mexico to New Orleans, from there up the Mississippi on a large steamboat to Memphis, Tenn., and from there by railroad to Knoxville, and from there to Tazwell by stage. Anyone who ever traveled by coach knows what that is, and there we were met by private conveyance, and carried to my grandfathers house.

PIONEER DAYS IN THE SOUTHWEST

O, what a happy meeting for father and mother and brother and sister to meet their loved child and sister after such a long absence away out in Texas. I remember how the relatives and friends used to come in to see my mother and hear her tell of her far away home, and how people did there, etc. After six months of enjoyment with relatives and friends we had to say good bye and return home.

My father bought cattle and moved to the frontier, Palo Pinto county, where we experienced all the dangers, privations and hardships of a frontier life. We settled on the west side of the Brazos river, one mile from Tom Pollard's. My father and hired man made two small cabins in which we lived through the winter. There was no lumber to be had there, so we lived on dirt floors, and cooked, ate and slept all in the same room. Next year they cut and hewed large logs and made a nice large room.

One morning my brother, Valentine, went out to get the calves, and not thinking about the Indians being near was not watching, when somebody said, "how!" and he looked and there stood an Indian with a rope on his arm and holding the other out to shake hands. My brother said he didn't make any effort to capture or hurt him, but he said that he, my brother, ran faster than he ever did in his life until he reached the house. He told us about it and we saw the Indian going across the bluff to go over in the other valley.

When the Indians were coming in after horses they hardly ever killed anybody, because if they did they knew the people would get after them and perhaps kill some of them before they could get any horses to get out of the country on. After they got all the horses they wanted, they would kill all who happened in their way or capture them, and very often would attack houses and carry off women and children and keep them to exchange for blankets, horses or anything they wanted. People were always on the alert, and watching for the red men. If we children went to the spring to get a bucket of water, we watched all the time to see if an Indian came out of the bushes or from behind a tree. We lived in constant dread and fear of being killed. If the dogs barked, we thought of Indians at once.

I remember one night we saw a light over in the mesquite flat or valley. Of course we thought it was the Indians camped over there, so my father said he would slip over there and see who it was. There was

3

a deep, dark ravine between the house and where the light was. He said he could creep up close without their seeing him and he could find out whether they were Indians, so he went on and got close enough to discover it was white people, he went on to the camp then, and it was Parson Slaughter and his boys out cow hunting and had camped there for the night. My father was a great hand to talk on the scriptures and politics and of course was hungry for a conversation and was soon so interested he took no notice of time nor of how uneasy we might be at home and stayed till midnight. I suppose of course that my mother thought the Indians had killed him and would come and murder the whole family. I was a little girl but I remember the agonizing fear we had. Nobody but those who have had the same experience can imagine how we felt. After awhile we shut the doors (if we had any shutters), and started out to go down a ravine and cross the river and go to Wesley Baker's, when we heard somebody talking. My mother was sure it was the Indians coming to murder us all, my brother being braver than the rest, said maybe it was our father, we waited a little while and heard our father's voice and some one with him, talking. Oh! what a relief to us all to know it was not the Indians. Parson Slaughter had come over home with father to stay all night. He was very much ashamed for having caused us so much uneasiness. Thus the reader can understand what the early settlers had to endure.

A neighbor, William Eubank, who lived three or four miles from us, across the river, enclosed his house and yard with tall pickets as a protection against the Indians. One day the men folks were all away, perhaps in the field at work, when some of the women or children discovered a party of Indians approaching the house, so they quickly set a bench by the wall and Miss Mary Eubank put a man's hat on her head and got up on the bench so her hat could be seen by the Indians, then she pointed a gun at them, when they saw the gun they halted, held a consultation, then turned and went away without molesting them. Perhaps she saved the whole family by her brave act; (she afterwards became our stepmother). Many other women have been as brave and defended their families from being butchered by the savages.

I think the Indians were on a reservation about this time and were friendly with the white people; some of them came down on the river to fish and hunt. They were camped near the town of Golconda, as it was called then (Palo Pinto now), and had their families with them. Pete Garland and several other men were drunk, and not having a very friendly feeling toward the Indians they shot in among them and killed several of their women and children. Of course this was wrong and insulted and enraged the Indians and they got on the warpath and the people had to "fort" up. My father was not at home at the time. The news came to us in the night and the hired man and my brother had to take us across the river to a settlement. The river was rising and the wagon got stuck in the quick sand, and they had to get out in the water and lift the wheels out. We went on to Ray Pollard's and stayed quite a while, then went down to Weatherford till the Indians were quieted down and had returned to the reservation.

My father then bought the Bob Dillingham place, a mile or two from John Pollard's, and we moved there.

My father often went to Weatherford on business and didn't return till next day. I remember one time he went and didn't intend coming home that night, but he said when night came on he felt impressed that he must go home, still he lingered but finally felt that he must go. We were living in a two room log house with a large hall between and my oldest brother slept out there. He had tied a horse in the field close to the house so the Indians would not get him. When my father was coming round the corner of the field he saw a man leading the horse out, and called to him to halt; the man didn't stop, so my father shot at him; he dropped the rope and ran. It was about one o'clock and of course we were not expecting him at that time of night so were greatly alarmed to hear his voice and the shooting; he rode around and explained. The next morning he went and got some of the neighbors to go with him and see if they, could find anybody, but I don't think they ever found out whether the thief was a white man or an Indian.

My mother said she had suffered a thousand deaths at that place for fear the Indians would come and kill us or carry off some of the children. Why men would take their families out in such danger I can't understand. We not only had the Indians to contend with, but the river as well. We lived close to the river and got water from a spring down under the bank, and whenever the river would raise three or four feet, it would get over the spring, then we would have to use river water.

My father kept a canoe to set people across the river. One time my father went to town, he generally went every Saturday. The ford was about half a mile from the house. We hadn't had any rain and the river was very low and we didn't think of a rise coming down, but just about the time he had gotten to the ford we looked out and the river was bank full and great banks of drift and foam coming down. We were very uneasy for fear he had got drowned, we went down and were standing watching the drift and wandering if he had got across safe, when he rode up to us. He said when he had got nearly across, he stopped to let his horse drink and was looking down and noticed the clear water rising up the horse's legs, he watched it till the water reached his horse's knees, then he looked up and there was a bank of water twelve feet high just above him. He spurred his horse and got out just in time to save himself, he ran his horse to the next crossing below and headed the rise and came back home.

My father built a little log house on a knoll near the river for my mother to teach school in. The floor was the ground, and logs split open and holes bored in the round side of the log and legs put in were the benches. We had a plank along one side of the room for a desk and a big crack in the wall gave us light; we had preaching also by whoever would preach for us. I remember two preachers who would preach for us sometimes, a Baptist by the name of Myers and a Brother Carpenter, who was a Methodist preacher. Our home was always a home for the preacher of any denomination; we loved to have them with us and minister to their comforts.

Most of the people lived across the river from us and children couldn't come to school regular on account of the river. So my father bought a place in town and moved there so we children could go to school.

In 1859 and 1860, I think it was, my mother assisted Mr. J. H. Baker in teaching the school in the little old school house on the hill side, on the north side of the creek, near Parson Slaughter's house. How well I remember the dictionary class. It was a large class of boys and girls, when we stood up to spell, the class extended around one side of the room and across the end. It was customary then when one got a head mark to go to the foot next time, and give the next one a chance to get a head mark. Lum Slaughter (now Col. C. C. Slaughter), was generally head or near the head of the class, and we would all laugh when he had to go to the foot, because he was the tallest one in the class, but he didn't stay at the foot long, he would soon go up head again, and he is ahead yet. I was at his and Cynthia Jewell's wedding. The dinner was set in the school house. There were other good spellers also, Molly Billy Hunty, Sallie Jewell, Will Covington and Cynthia Jewell, who were always near the head of the class waiting for their time to get a head mark. There were others that I can't remember.

Well, the war came up then and our mother died, her father had lots of slaves and she was raised very tenderly, never having done any work before she was married. The hardships and continuous fear of a frontier life was too much for her. I remember her father came to Texas once to (see her; and preached in the school house where she taught, he was not very favorably impressed with the country and didn't stay very long in Texas. After our mother died we children had to learn to card and spin all the cloth our clothes were made out of. We used bark and leaves from oak and walnut to color the thread with; walnut leaves made such a pretty dark brown and broom weeds made a pretty yellow, we used moss and other things to color with. We had to do without lots of conveniences and necessities; for tumblers we would take a long, smooth bottle either black or clear and take a buckskin string and see-saw it around the bottle till it got hot then drop a little cold water on it and it would come apart; we were quite proud of them to drink milk out of. For forks they would twist wire together. For soda we would boil weak lye down to a potash after we had made a pot of soap, it was a poor makeshift but better than nothing. We parched wheat for coffee. I remember we gave $30 for a pair of cotton cards and $5 a yard for calico, very poor calico at that. We didn't use hardly anything that was not homemade. We used to sit up till ten or eleven o'clock carding, spinning or knitting.

We didn't have any near relatives in the war but some of our neighbors did; and how hard they would have to work to make clothes and blankets and socks for them.

PIONEER DAYS IN THE SOUTHWEST

We thought we had a hard time on the frontier, but it wasn't anything to compare with what they had to suffer back in the old states. My grandfather said he was unmercifully beaten several times and all of his valuable property taken. My husband's sister had her baby's cradle set out in the yard and the men told her to get out, they then stuck the broom in the fire and got it ablaze then stuck it to the feather bed and burned up her house and everything she had and left her and her children standing there helpless with nothing to eat and nowhere to sleep. They were Jayhawkers that did this. In Tennessee another sister-in-law looked out one morning and saw the blue coats (as the Union soldiers were called) everywhere, and all around her premises. She went to the commander and told him she was a widow, that her son was in the army, and that she wanted to be protected. He said: "Go home madam, and rest perfectly easy, not a thing you have shall be molested." And not a chicken or a fence rail was touched.

My father was assessor and collector during the war and when his term of office expired he didn't want to go to the war and leave us unprotected so we moved our cattle to Menard county and settled at the Bowie Spring where the former James Bowie had a fight with the Indians, five miles from Menardville and three miles from the old Spanish fort.

My father found a sabre on the battle ground and made a graining knife out of it. What is that? some of you young people say. It is a long knife to take the hair and grain from the deer hide; when they want to dress one they then put the hide in a strong suds made of lye soap and water and a spoonful of lard; they let it remain all night by the fire to keep it warm, then would wring it out as dry as possible, then pull, stretch and rub it till perfectly dry, and soft as cloth, then it is ready to make gloves or pants. I have made many a pair of gloves and pants out of the dressed skins, and I made a shirt out of a fawn skin and colored it yellow with chaperal root. It was very nice and durable to wear over the other shirts when running through the brush after cattle.

Cellery spring was up the valley a mile from where we lived; it ran about a hundred yards and sank in the bed of the creek, then came on down and made Bowie spring, it ran out of a cave, the water was about a foot deep and very clear and sparkling; it ran on down to the

Sansaba river near the old Spanish fort. Tradition has it that Spaniards used to live in that part and worked the mines that are in that part of the country and the Indians attacked them and they buried their money and the Indians killed them all. Different parties have explored the country and tried to find the treasure but were unsuccessful.

I think we moved to Menard county the fall of 1864. The Indians didn't bother people much then, but later got to be troublesome. They came to Fort McKavitt on the head of the Sansaba river and killed one man and stuck a spear in a girl, she pulled the spear out herself. Then they gathered up a large bunch of cattle, hundreds of them, and drove them off. The men followed them as soon as they could get together but could never overtake them.

My husband taught the first school ever taught in Menard county at Fort McKavitt. Then he taught in Menardville. Then we moved back to the ranch. One evening my husband was busy at something out in the yard; there was a large thicket north of the house about two hundred yards; we had a good watch dog, and he barked toward the thicket and would sniff and seem uneasy all the evening. My husband didn't say anything about it at the time but about twelve o'clock in the night (the door was not closed) he saw the dog slip by the door squatting low on the ground. He knew at once there was something wrong, so he jumped up and ran to the door and there was an Indian on one of the horses right at the end of the porch; he ran back to get his pistol, which was hanging on the bed post but the Indian saw him when he first ran to the door and ran away. Then we could hear them passing down on each side of the spring branch, not more than fifty yards from the house. They got all the horses on the ranch but one, and were so elated over their success that they went over the hill in a little valley and had a war dance, we could hear them very plainly whooping and yelling. My brother in-law wanted to go down the branch and slip up the ravine and shoot some of them, but my husband wouldn't let him do that, he said they might come and burn the house and kill or capture us all and carry us off.

I can't describe my feelings. I had never heard an Indian yell before. There were only two women, two babies and two men of us, and I have no idea how many Indians, fifteen or twenty I suppose, and we being out on a ranch five miles from Menardville it was too much for us. For fear they would soon come back again and be more bold, we decided we'd better abandon the ranch. We packed up and left. We went down to the town of Burnett where my husband worked in the store for John Alexander one year, then we returned to Menardville again.

My father gathered up all his home cattle and started to Mexico to sell them. Before he got to the Pecos river, however, a much larger herd overtook them and they all went on together till they got to the Horse Head crossing on the Pecos, when a large party of Indians (the Apaches, I think) came on them and surrounded them. They thought they'd starve them out and make them perish for water, but some of the men slipped down to the river and got water. I think they kept them there three days, then decided they would drive the cattle off, so drove the entire herd off and we have never got any pay from the government for them yet.

Down in Llano county a man lived named Bradford; a woman whose name was Mrs. Friend was visiting the family when the Indians attacked the house. It seems the Bradford family got away somehow and Mrs. Friend had to defend herself the best she could. They said she fought the Indians with a smoothing iron, and they left her for dead. By the time they left she was weak from loss of blood, but she raised up and pulled a quilt from off the bed and laid her head on it. In a short time the Indians returned, but she laid just like she was dead so they departed. The next morning when Bradford returned home she was sitting up smoking. I could relate many more incidents of Indian depredations, but can't remember them clearly. My memory is not good, I can't remember dates and if I have not stated these facts just as they were I hope the reader will excuse my mistakes. I relate them as I remember them from childhood partly. My children are all grown now. My four girls were school teachers but are all married now. My oldest son served six years in the United States army; my youngest son is at home with me.

I think there should be a reunion of all the old pioneers in the near future. I hope we will all be at the grand reunion in the sweet bye and bye. I have many loved ones over there waiting and watching for me.

PIONEER DAYS IN THE SOUTHWEST

PIONEER DAYS IN THE SOUTHWEST

CHAPTER X. BY TILATHA WILSON ENGLISH, GAINESVILLE, TEXAS.

Dear old-timers, as I am an old Texan and an old-timer, too, I wish to step in and say a few words on old times in Texas and how the people lived. My father left Kentucky when I was two years old, went to Illinois and stayed there six years then started to Texas and got to Arkansas and some of the family took sick and we stayed. Father got a place and we stayed there three years. My oldest sister married while we lived there and we all started to Texas together. My father had three yoke of oxen and my brother in-law had two yoke. We came to Fannin county just at Christmas. In a short time our oxen began to die. They all died except one of my father's so he got a job for himself and my brothers making rails and they soon paid for another ox. The day they got through paying for him he died and we were left again.

Father rented about ten acres of land and made a "half-yoke" he called it and worked that old ox, plowed him like a horse and tended his crop. He raised plenty of corn to do him. They never raised cotton in those days; only a little for home use, and they would work that old ox and ride him anywhere they wanted to go. When father got his crop laid by he got a job digging wells and was able to get another ox and a cow and calf. Then he gathered his crop and raised peas enough for us and all the neighbors. He heard about land coming in for settlers in Grayson county, so he came to Grayson in 1845 and took up 640 acres of land.

The first year we came to Grayson, my father bought a prairie team, which was six yoke of oxen and a big prairie plow, all on time and my two oldest brothers began breaking prairie; they broke a small field then began to break land for the man to pay for the team. Father bought his bacon that year, and what kind of meat it was! It was bear bacon and it was real good. There was no mill nearer than Bonham, so most everybody got a little steel mill and ground their corn at home, sifted out the fine and used it for bread and cooked the rest for hominy. They did not raise any wheat for a few years, but when people got

12

their land in cultivation they raised lots of wheat but hardly knew what money was. Everybody was poor who came to get homes.

When they wanted a rope if they had a rawhide they made it out of that, but if they did not have it they would spin coarse thread and make rope out of cotton. We had rope works we made them on. I have spun and helped make many a rope and bed cord.

My father had a little shop and when they could not work in the field, they worked in the shop. He made all our shoes and made looms for the women. All the way we had to go to church when there was any, was to walk or go in an ox wagon, or on a slide. Some had no wagon, and they would make a big slide and put a floor in the bottom, and cross-pieces to sit on, and hitch a yoke of oxen to it and go wherever they pleased. There was not much preaching those days to go to, but when we would hear of preaching any where near, we would go. I have walked five miles to church and back the same day, and wore homemade dresses, bonnets and shoes. My father sometimes made buckskin and sometimes cloth, and sometimes leather shoes— just anything he could get to make them out of, and we would be mighty proud of them, and most everybody dressed alike. People loved each other then; it was not as it is now. Pride and style have taken the place of religion. I am an old spinner and weaver. I have had to quilt, card, spin and weave ever since I was large enough. We had to pick out all the seed from our cotton with our fingers for several years after we came to Texas, make up our own clothing, and raise our own indigo to color the cloth. I never saw a sewing machine or cook stove until after the war. I was married, and had three children before I ever saw a cook stove and in making up our cloth after we got it, we would sew it all with our fingers and make the buttons out of thread. I have made many a button out of thread. You think, no doubt, that we had a hard time, and so I reckon we did, but we enjoyed it, and it was happiness to us all, for we loved each other, and were always glad to see and help each other and when one got sick we would go eight or ten miles to see them or sit up with them, if it was needed.

I know some that say they would not like such times as we old timers had, but if I was young again I would love to be in such a place and would go to it, if there was such a place.

PIONEER DAYS IN THE SOUTHWEST

The only thing that I would not like would be no schools and churches, and as I am such a lover of church, I could not live where there was no church, for all the real enjoyment I have is going to church.

I have merely given a little sketch of our early days in Texas and now about the war times. As for myself, I had no harder time in time of the war than I had before and after the war. My husband was not in it, but was freighting for the government, hauling cotton down in southern Texas. The longest time he was gone was fifteen months. My neighbors were all good to me, and they chopped and hauled my wood, and my two oldest little boys were large enough to carry it in, go to mill, feed the hogs, and lots of other things. I had to spin and weave all our clothes as I had always done, but I heard of so many who had worse times, it made me feel sorry for them, and I felt the Lord had blessed me and I was doing well.

I will tell you what kind of furniture we had when I went to keeping house. When I married, our bedsteads had only one leg. We hewed out a square stick and bored a big hole in it, and one in the log of the house and put in a pole for a side rail, and one in the leg, and the log for an end rail, and one across the back, laid boards on that and our bedstead was done. Our table was four sticks with small boards around to make the frame; then boards nailed on for a top, if we had nails; if not, they were put on with little wooden pegs.

Our china closet was holes bored in the logs of the house, pins put in them, a board laid on that, and another one above that and so on until we had as many shelves as we needed. Clothes shelves were made in the same way. Our chairs were wheels sawed off the end of a log, the bark taken off, and one side basined out a little and legs put in then, a wide board put in the back and one chair was done. This was the principal part of our furniture.

We raised gourds to use for almost everything; big gourds (we called them fat gourds). Lots of them would hold a half bushel. We put lard in them or anything we wished. Spanish gourds were large at both ends and small in the middle. We would saw off both ends about half way, clean them out good, tie a cloth over one end, and it made a good strainer. We had gourds to milk in, drink water out of, and other things too numerous to mention.

14

PIONEER DAYS IN THE SOUTHWEST

When our fire would go out we had to go to a neighbor and get fire or catch it some way. I have many times taken a skillet lid and an old case knife, and knocked fire out of it. I have spun fire many a time, and I will tell you how I did it. I took deep copper thread and doubled it several times, and twisted it a little, then held it in the whirl of the wheel and turned the wheel right fast, and it would mighty quick set it on fire. I would have some cotton ready and stick in it. I would soon have plenty of fire.

I have heard of people weaving gallouses with button holes already in them. I have woven tape for gallouses, bridle reins and saddle girths, but I never learned to weave holes in them. I used to knit my folks gallouses and knit the button holes in them. I used to plait wheat straw and make my men folks summer hats, and take home-made jeans and make them caps for winter. In time of war I plaited wheat straw and made a Baptist preacher a hat. He was proud of it and kept it to wear to church. He was a good old man and gone, long years ago, to his happy home. He left Grayson and went to Tom Green county and died there; his name was Tommie Cotton. Perhaps some of you old timers may have known him.

When we came to Texas in 1845 the country was full of all sorts of wild game—turkeys, deer, bear, panthers, antelopes, wild cats and almost anything.

Alexander Wilson, who was my husband, was a great hunter and killed lots of game. He almost kept the neighbors in meat. He has brought us many a good ham of venison long before he and I thought of marrying. We went to school together; all the school I ever went to, and it was an eight months school. I went half the time. My mother was dead and my sister and I who was three years younger than myself, had to keep house for my father and five brothers, so we had to stay at home week about and keep house and spin. Our school house was a log house with a dirt floor and a big wide fireplace. Our seats were logs split open and legs put in them, the flat sides up. That was all the school I ever got to go to, and what little I know I learned myself at home. So you see why I don't write and spell good.

There were plenty of bees in the woods in those days. People could go out and find a bee tree any time they tried and cut it down and get lots of honey, and save the bees if they wished, so we soon had plenty of bees and those days we raised lots of pumpkins and that was a big part of the people's living. We had no flour for several years, so we would stew our pumpkins till done and put it in meal and salt it, if we chanced to have salt, and work it up into dough, making it into small thin cakes called pumpkin bread, When we had hog meat we would fry a few pieces, take the grease and crumble corn bread in it, putting in water and salt, and we had a pot of soup called "poor doo." We thought it very good those days. People think they live hard now, but they don't know anything about hard times. Most everybody lived the same. We have sat down to eat many times and had nothing to eat but cornbread and meal coffee, but we were satisfied, and would work away trying to get a start, and so we did.

PIONEER DAYS IN THE SOUTHWEST

PIONEER DAYS IN THE SOUTHWEST

CHAPTER XI. BY GEO. ELY, ODESSA, TEXAS.

I was born on Sept. 25th, 1840, in what was then Marion county, Arkansas, and my parents were too poor to give me an education, and so I grew up without, except such as I could pick up by reading books, papers, etc., so that when requested to write a chapter for a book, I felt at a great loss to know how to proceed and especially as my memory has failed to such an extent that it is often impossible for me to recall events accurately.

My father emigrated in 1852 to Texas, settling a homestead in the eastern edge of the "Little Cross Timbers" six miles west of Black Jack grove in Hunt county (now called Cumby I believe). We got our corn ground into meal by a windmill that stood on an eminence a half mile west of Black Jack grove, which was operated by a Mr. Bowerman I think; there were no steam mills then in that county, but they came a few years later.

People got most of their corn and wheat ground on mills run by inclined wheels turned by the weight of four or five oxen walking as the wheels turned. These mills did a fine service in their day and became more numerous as the demand grew, until the steam mill was introduced which soon put the inclined wheel into "the things that were." Speaking of mills reminds me of a very ancient machine that was used by the early settlers for converting corn into meal; I allude to the "Show Tom." This contrivance was made by securely fastening a water trough to one end of a long stiff pole; to the other end a "pestle" was firmly fixed so that its lower end would just drop down into the bottom of a basin dug out in the top of a stump, or a large solid rock; the pole being fixed on a pivot in such a way that when water was run into the trough that end down and the pestle up from the corn that was in the mortar —this would let the water out of the trough and up it would go causing the pestle at the other end of the pole to drop down into the mortar *on* the corn. The process went on and the "Tom" could be heard all through the night; and so there would be meal for breakfast unless coons or bears got to it first.

From Hunt county westward the country was very thinly populated and Fort Worth was then very far out west. Game of all kinds abounded; and very few people depended on farming, as it was

18

then too dry to raise much produce, aside from wheat and a little corn, etc. About 1867, I went with my uncle, J. J. Keith, to aid him in removing his cattle to Erath county, which had began to settle up, but there were only a very few ranches west of there, and deer, turkey, and antelope were very plentiful all over the country while the grass was as fine as the heart could wish, and stock would keep rolling fat all the year round. It was easy raising stock then, but the Comanche and Kiowa Indians and others became very troublesome and would make their raids into the settlements, generally in the light of the moon and take what stock they could get their hands on, and often kill and scalp whoever came in their way, frequently carrying women and children into captivity. Thus was Cynthia Ann Parker captured while young, of whom so much has been said and written. These Indians became very bold and defiant, and a man never knew when leaving his home, whether he would ever return alive, or supposing he did, whether he would find his family, his loved ones, at home or in captivity. People would go five, twenty, even thirty miles to borrow meal or flour and do neighboring generally. They could do no better, as they were so few of them, except in the little villages, and they had to do so. When their little crop of grain was to be ground, they would go thirty, sixty or eighty miles to the mill, and as there were no wire fences the cattle were turned loose usually where water was, and allowed to roam at will. This necessitated much hard riding during the branding season, for the cattle would scatter over several counties. The men usually went in squads of five to ten or more together, and when they rounded up their cattle at the nearest branding pen, they would help each other, and when they found a "flitter ear" as they called those yearlings they had missed the last season, they each knew about how many they were entitled to, and so would apportion them out among themselves to their mutual satisfaction. Very seldom did they fail to be mutually satisfied with results, as those frontiersmen were a noble class of men, living for each other, and co-operating with each other to a remarkable degree. So occasionally they would round up a few cows with calves that needed branding and no owner there to look after it. Well, the boys would look for the brand of its mother and brand it accordingly. Thus everybody helped everybody else. But it was awful the way they lived in constant danger from the murderous

Indians. Were they afraid? Well, if they were not, it was because they had become so accustomed to the danger that it hardened them against fear, and those frontiersmen were a venturesome class of men who knew no fear; and it was necessary for the way to be opened up for the occupancy of the country by an industrious people.

The commerce of the country was carried almost entirely by ox wagons, and two to six or seven yoke constituted the teams. These teams would be driven hundreds of miles to Houston, "Jefferson, Shreveport and other places, for dry goods and groceries, or to the piney woods of east Texas for their lumber. This would occupy weeks together for a trip. Common boxing lumber was worth five to six dollars per hundred feet at Stevensville. No railroads in the state except fifty miles of the H. & T. C. Then men that did the hauling were called "wagoners," and a very useful class they were. I remember there were very few bridges then, and so they had to ford the streams they crossed, which was often very hard to do. In wet weather especially was this difficult. It was no uncommon thing for the heavy loaded wagons to sink to the hubs in mud, then it took work to get them out and on firm ground again; this was accomplished by putting two or three teams to the wagon and pulling it out by force, as generally two or three wagons travelled together. Sometimes they dug the wheels out, placing pieces of wood under them till they could be pulled out. Again they would be compelled to unload and carry the stuff to firm ground, then draw their wagon out. Thus it sometimes happened that a man would work all day in a piece of swampy land and camp at night, then return to where he had camped the night previous for fire brands with which to start up his fire for the night. This will give some idea of the hardships of the early settlers, what they had to contend with, and overcome before west Texas was really ready for civilization. There is another thing that must not be overlooked; I mean the kind of foe the frontiersmen had to contend with and drive out in order to open the way to civilization. The Indians, be it remembered, studied and practiced war and plunder, were trained to it from their infancy, so that to be able to take many scalps and horses was their chief delight; while the settler was not trained, but had his domestic affairs to attend to. This put him at a great disadvantage; it was not infrequently the case that Indians proved even more than a match. Here is a case to illustrate:

Six or eight men happened to come upon a lone Indian and gave fight, but that Indian was on to his job, and kept his arrows flying after each other thick and fast; meantime constantly changing his position jumping about equal to a jumping jack, and in spite of all the men could do, he got away. The boys just could not hit him. Very few people now can know half of what our brave bordermen had to contend with. Then again we must not forget that our people only had muzzle loading rifles and pistols to fight with. Metallic cartridges had not yet come into use. The Indians had a shield made of thickened raw hide which took the very best rifles to puncture, and they could nearly always manage to keep them a little sidewise to the bullet, so that instead of penetrating the shield, it would glance off harmlessly.

About the year 1858 or '59 the Caddos, reserve Indians, in what is now Young county, and the Kickapoos and Tonkawas, higher up the Brazos became very troublesome, stealing and killing stock, and sometimes killing a man, although the government had them on their reservations and feeding them. On one occasion in or near the Keechi valley, on the Brazos river, a squad of our men had a fight with the Indians, in which Sam Stephens was killed. He was carried to Golcondo (Palo Pinto) and put in his coffin, and laid out over night in a wagon; meantime a runner was dispatched to his father at Stephensville with the sad news. His father, John M. Stephens, quickly gathered up a squad of men and hastened to him. The grief stricken father climbed up and looked over the side of the wagon, and with just a glance at the casket containing his beloved boy, came very near falling. We buried Sam in a temporary grave about a half mile west of the little village in a nice cave, firing a volley over his grave, thus burying him with the honors of war. These reserve Indians were an eyesore to the Texas frontier, and the time soon came when the people began to make preparations to wage a war of extermination against them. Men began to assemble within a few miles of the reservation by the hundreds under different leaders, as Captain Peter Garland, John M. Stephens, Captain Cook, Captain Baylor and others, there being two or three divisions or camps of the men. One party had a fight with the rascals, but I have forgotten the results. About this time the governor took a hand and called for a company of enlisted men to aid in removing both reservations of Indians across the Red river into the Indian Territory. This company was placed under command of Captain John Henry Brown, the writer being a member of it. Early in July 1859 we went into camp near an old Indian village, ten or twelve miles from the Caddo Indian agency, and on the bank of the Brazos. I was hardly grown and carried a brass barreled pistol, and so the boys called me "Brown's artillery."

When the Indians finally moved we followed them to the Little Wichita river where we remained for quite awhile, and where a lad by the name of Estes and myself, were stricken with dysentery. Poor Estes died, and I came near dying. They buried the poor fellow far from human settlements and without a coffin. From there we returned to our old camp on the Brazos, and here those who enlisted with me at Stephensville, received our discharges by the favor of Captain Brown.

It was on this trip that I first saw the Mexican cart, two-wheeled vehicle, drawn by two or three yoke of oxen, their yokes being placed on top of their heads, and lashed to and just behind their horns with strips of rawhide. The wheels of these carts which hauled our camp supplies, were some six or seven feet high, and their tires, five or six inches wide. We had, I think, some fifteen or twenty of these carts, and their drivers were a merry set of Mexicans, who spent the most of their leisure time playing "monte," while at night they usually indulged in songs and jests until late hours.

I now pass over several years, for the great war soon come on. I married Miss Letta B. Eatherly in Franklin county, Arkansas, and finally located in Eastland county, near the Mausker lake. The Confederate government required all the able bodied men between the ages of eighteen and fortyfive years to be enlisted for local defense of frontier settlements against the Comanche and other hostile tribes. So in April, 1863, a company was organized and mustered in at the Mausker ranch —composed of the able bodied men in Eastland, Shackleford and Calahan counties, with four men from Comanche county. After enlisting all those, we had only forty men; this will serve to show how thinly the country was settled.

PIONEER DAYS IN THE SOUTHWEST

Our company was divided into four squads or "scouts" as they were called. One scout would take their "grub" etc., and go out generally to the westward of their homes and look for Indian signs and do general scouting for ten days, when they would return to a designated place to meet and be relieved by the next "scout." Thus we served ten days and attended our own affairs thirty days.

Sing Gilbert was our first captain—a noble and brave man. This service was continued for something like a year without any very exciting occurrences in our company until in August 1864. My time to go out came near, but it so happened that for some cause, my cousin, T. E. Keith (who is at this writing the county attorney-elect of Stephens county), came to me and asked to take my place for this trip or scout and as he got mixed up very badly with the Indians in a terrible fight I will let him tell of this scout and its results.

On the 8th day of August, 1864, this scout of seven men started out under James L. Head, a corporal. We went west to McGaugh Springs and camped for the night; on the 9th we started west intending to follow the Leon river to its "head." (I think he means the south fork of Leon). Not more than twenty miles, but we got only five of six miles when we discovered the trail of a large band of Indians coming down towards the settlements. This trail was discovered near where the Texas Central railroad crosses the Leon valley, seven miles east of Cisco. This little band of men, poorly armed, took up the trail of forty or more Indians, and followed it southeast, some twenty-five or thirty miles, to a place where the town of Jewell now stands, overtook them, and immediately engaged in battle, but the Indians proved to be ready for fight, so Corporal Head soon ordered a retreat for the Gilbert ranches, three miles away. About two-thirds of the Indians were afoot, the balance riding broken-down Indian ponies which they expected to exchange for "White man's horses. Well our little band of men afterward scattered to the two Gilbert ranches, which were one and a half miles apart, for recruits. Captain Sing Gilbert lived at the lower ranch. They all at once got their guns, mounted and met at the upper ranch for a further run and fight, and we got it. There were then twelve men, with Captain Gilbert in command. We went back to where the fight had occurred, took up the trail and followed it twelve or fifteen miles to what is known as Ellison Springs, where we overtook the

Indians for the second time that day, twelve against forty or more Indians and two-thirds of them on foot. Captain Gilbert then and there without any formation of his men, ordered a, charge and led right up to, within thirty feet of those foot Indians, halted and fired, then ordered his men back. Well, he paid for his indiscretion with his life. An arrow struck him in the neck as he turned, and in an hour he bled to death. Button Keith's horse fell and he was killed on the spot. Tom Gilbert, Tom Cadenhead and Jim Ellison were all severely wounded. So there was nothing left for ITS to do but to run for our lives to Ellison's house three hundred yards away. Five men out of twelve killed or disabled, pretty severe fighting. If any Indians were killed they carried them off as was their custom.

To show the endurance of those Indians, will say that this fight occurred about four o'clock p. m. They had already traveled thirty-five or forty miles that day, had two fights; then went to Stephensville in Erath county, thirty-five or forty miles further and stole fifty or sixty head of horses, before daylight next morning.

Our two fallen braves were buried at Stephensville, thirty-five miles away, the nearest burying place. This disastrous fight will illustrate in part what the frontiersmen of Texas had to contend with. We had poor arms, the muzzle loading rifle and a few old style "Cap and Ball" Colt's pistols and the home-made single barrel pistol; usually made from an old rifle barrel, and by local gunsmiths, converted into a very fair pistol and reliable for one shot. Then the powder we used was all home-made, as we were cut off by the war blockades from commerce and much needed supplies. This powder would kill, but was not reliable, and our boys called it "slow-push" powder. Our gun caps were also home made, were of lead and very hard to explode.

These frontiersmen seldom stopped to "count noses" as this fight illustrates, but went in like a whirlwind when duty called and with such arms and other facilities as they could snatch up. These heroes are nearly all gone and will not again be needed as frontiersmen in Texas, but a braver, nobler, truer set of men never lived. He was frequently dressed in his buckskin and was rough and ruddy in his exterior, but in his heart he was endowed with a noble and unselfish disposition. What he had was yours, if you needed it, and what he could do for you in your distress he would do with a will. God surely put him here for his special but dangerous work, and but for him, Texas today would no doubt be the home of the red men and buffalo.

It seems almost like magic to me; often as I ride in a fine passenger coach over our railroads and see the country thickly populated with an industrious, enterprising and happy people, and the country dotted over with towns and cities where forty or fifty years ago I used to see nothing at all that represented human habitation, and it does give me a tinge of satisfaction that in my small way I did what I could to make this glorious "Garden of the world," habitable for a people worthy of it.

I will be sixty-eight years old just one month from today, and doubtless ere long will be buried in the ground I fought for. I pray God that this land of heroes may become as loyal to Him as they are to their Texas homes.

PIONEER DAYS IN THE SOUTHWEST

CHAPTER XII. BY S. P. ELKINS, TISHOMINGO, OKLA.

IN the fall of 1870, Captain J. M. Swisher raised a ranger company for frontier protection; I joined his company and went west; we were stationed in Coleman County, Texas, at camp Colorado, and in the winter moved down on Home creek. We had a picket station of fifteen men at the mouth of Concho creek in Concho county. There was one ranch about twelve miles from the picket station; this ranch was owned by Rich Coffey. He kept a large force of men hired for protection, as he was not near any settlement, and there was nothing in that country but wild cattle, buffalo, mustang horses, wolves and a few panthers, and the Kiowa and Comanche Indians often made raids through there, then would go farther east where the ranches were closer, and there steal horses and kill all they could.

Those that were on the frontier had much to endure. They did not know at what time they were going to be killed by the Indians, so they had to do the best they could. The ranches would have stockades built around the houses and would have port holes cut on all sides of the house so when the Indians attacked them they could protect themselves and their families.

The people of those days had something to think of. There were no neighbors near to lend a helping hand. Some places where the neighbors were close enough they would have preaching, maybe once a month. Everybody went armed all the time. The men wore their pistols the same as their clothes. They would take their families and go to meeting, take their guns along and stack them in one corner of the house until after meeting. They were glad to see each other and would shake hands when they met, and also when they parted, thinking maybe for the last time. When a stranger came about he was welcomed in and made to feel at home, no charges, glad to see any one.

When a man left his family he didn't know whether he would find them alive when he returned or not.

There was a family killed in 1870 in Brown county, I have forgotten the name; the man was in the woods making rails, when he

26

heard his family screaming, and started to them, and saw the house surrounded by the Indians. He had to stand and hear them scream their last screams, he having no arms with him, as he thought of no danger on leaving home. The whole family were killed, the children's brains knocked out. The news came to our camp and we started out after them and about fifteen or twenty miles from camp we struck the trail of the horses and followed them as fast as possible until we reached a country where there were a great many wild mustang horses. Here we were bothered for some time; the scope of country that the horses roamed over was a large one and the weather being dry and the ground hard, made it difficult to trail the Indian's horses. Then we found where the buffaloes were traveling in great herds—we could see thousands at a time—that bothered the trailing of the Indian's horses, for at best it was difficult to trail them, the ground being hard and dry and covered with mesquite grass. We came to places where-they had camped on high points, and other places in going across mountains where they had pushed their horses off of high cliffs where we could not get our horses and pack mules down, and we would have to find a better way, and would lose time. On the sixth morning we struck lots of buffalo that bothered us again, so most of the scouts were walking and leading their horses, trying to trail the Indians. Some of the boys were off to one side shooting buffalo and they came close to one of the boys that was down walking; his horse became frightened and jerked loose from him, and the last that we saw of the horse he was going back toward home; the owner, Mr. Stover, followed him, we had no time to follow horses on the back track. The last we saw of Mr. Stover until we got back to camp, he was still following his horse. He was left about a hundred miles from camp, no houses or roads, nothing to eat except what he killed. We traveled all that day and camped that night, and the next morning while loading the pack mules and getting ready to start, the captain and one of the men went out to strike the trail, the captain had a field glass and was looking over the prairies, he saw the Indians about two miles distant. The captain sent this man back to tell the boys to come as soon as possible. The boys started ranger style; all get there that can. The boys were all strung out across the prairie. Two of the Indians were after a buffalo, and the balance were driving a bunch of horses. We were nearest to the two that were after the buffalo; they had shot the buffalo and were down cutting some beef from it, and did not discover us until we were about a quarter of a mile from them. They mounted their horses and started toward their comrades. The Indians that were driving the horses left them and began trying to get away; the riding commenced then. We had to run for several miles before catching them.

There were only three of us in the running fight; the boys were scattered three miles apart. The run and the fight were about fifteen miles long. The Indians ran into a cedar brake and got in a cave in the head of a ravine. We dismounted and went into the cedar brakes, as it was too rough to ride into. When the boys all got there, a part of them were sent on top of the mountain and the rest at the mouth of the canyon, then there was one man left with each bunch of horses, the balance of us went into the cedar brake after the Indians. We hadn't gone more than two hundred yards, looking every way, but couldn't see anything. All of a sudden the bullets began to pour down on us; but we could not see where they were coming from, so we left for a little while until we could reconnoiter and get them located. We got but two of them, got all the horses they had and all the blanket saddles; got but one man shot. Well, night came on and all of us in the brake and all worn out, no water and our horses tired out; we started back to our pack mules, about fifteen miles, where we had left them that morning.

We got there late at night and the next morning we gathered up all the horses we had got from the Indians and started back to camp. That fight was on the eleventh day of November, 1870, and on the thirteenth day it commenced snowing; snowed all day and all night. We awoke next morning nearly frozen to death, as we were in the prairie with no wood; so we saddled up and started, but the snow got so deep that we had to lay over two days. You could ride into snow five or six feet deep where it had drifted into ravines and gulches. We got on the south side of a mountain where there was some scrubby cedar, and there we stayed for two days and fared very well. We would tie the tops of the little cedars together and rake the snow from under them, then spread a blanket over them and build a fire. Then at night we would rake the fire out, spread down our blankets and do the best we could until morning.

PIONEER DAYS IN THE SOUTHWEST

When we got back to camp, the ranch men gathered in at Camp Colorado and we all had a time. We had some good times and some bad ones. There was plenty of game such as buffalo, antelope, deer and a few bear and panther, but it soon got monotonous, seeing them every day.

The first three months that we were out the government furnished us rations; our bread was hard tack, full of worms, weevil and spiders; we killed our own meat. The next three months our rations got misplaced and we did not have a bite of bread for twenty-two days; we went to Comanche Town, Texas, there we got one box of crackers and went back to Coleman county in the dead of winter. Captain Swisher and four of us started to Fort Mason for supplies, and the second day we met a train of government wagons with a company of United States soldiers bringing rations to us, then we were all right again.

I was discharged in 1871, and went, to Palo Pinto county, Texas, settled on some vacant land and started a hog ranch. I put in some new ground as I had to keep hands hired on account of the Indians. When we would go to work we would take our guns along and set them near us for fear of being attacked by Indians. One of my men was shot by some traveling men that stayed with me one night. They mistook him for an Indian. I saw that kind of business wouldn't do, so I quit and went back to the settlements.

In 1874, Captain Perry of Blanco county, Texas, was getting up a company to go on the frontier, and I joined his company and went to Menard county, Texas; our headquarters were near Menardville. We stayed there for some time but had no fighting to do.

There were six outlaws came to Menardville one day and shot up the town. The sheriff came to our camp and asked our captain for help. The captain detailed six men to go and capture them. We had no trouble capturing them as we found them all asleep.

In a short time we moved out on the Sansaba river near the Raganaw mountains. We kept a picket on the mountain and our horses in the valley so the picket could see the Indians if they attempted to make a run on us and steal our horses. One day the picket discovered a trail through the shinoak brush and followed it until he came to a cave; he heard something in the cave making a noise and he came to the camp and got permission to shoot a bear. Most of the boys went up the mountain to have some fun killing bears. The shinoak brush was very thick all over the side of the mountain and about as high as a man's head, and when we got up in front of the cave there was a clear spot; the boys had gathered here ready to shoot a bear when it was run out of the cave, One of the boys threw a rock into the cave and something would make a growling noise and grind its teeth; everyone was ready to shoot the bear when he ran out, another rock was thrown into the cave and in a moment out came an avaline or musk hog, and darted through the crowd, it was so quick done, gnashing his teeth as he came, that the boys all fell over in the brush and gave him room as the brush was too thick to run through. Their scare was soon over and no one hurt. There was another rock thrown into the cave and out came another avaline; he was killed, another stone was thrown, and out came the third avaline and he was killed and that ended the excitement.

I stayed with my company but a short time when I was detailed to go with Major John B. Jones as one of the escort; there were fortytwo of us in the escort. There were six companies of Rangers on the frontier, and seven men from each company were taken to make Major Jones' escort. We traveled all the time from one company to the other, the companies were about a hundred miles apart.

On the third day of July, 1874, the Lost Valley battle was fought; twenty-eight men under Major John B. Jones against between one hundred and fifty and two hundred Indians. We had thirteen horses killed and wounded, two men killed and two wounded. In November, in Menard county, we had another fight; killed six Indians and captured one. We took this Indian to Austin and he was sent from there to the state prison as a public enemy. The fight was a running fight of about fifteen or twenty miles. The Indians ran onto two of the boys that had gone out to kill beef; they came to camp and reported, the horses were run in as quick as possible and we were soon after the Indians.

Scott Cooley was in this fight. His people were all killed by the Indians several years before in Palo Pinto county, in Keechi valley. Scott at that time was a small boy and the Indians took: him captive; he was afterwards recaptured by the whites. This gave Cooley a great hatred for the Indians, and when he got a chance he fought them hard and close. Cooley had no fear and had a blood thirst for the Comanches.

In 1877, I moved to McCallow county; there were about one hundred and fifty inhabitants in the county and about one hundred of them lived in Brady City. This was a very dry year and everything was scarce. I hauled two loads of corn one hundred and ten miles and as soon as I got them home the people would want corn bread. They would get in wagons and go farther west to the buffalo range, and kill and dry meat for their families; but in 1879 the buffalo were all killed; there were about 1,100 hunters on the range at one time killing buffalo for their hides and tallow, but when the buffalo were killed, that killed the Indians.

PIONEER DAYS IN THE SOUTHWEST

CHAPTER XIII. BY JOHN A. LAFFERTY, PARKER COUNTY, TEXAS.

I have a stiff arm and can't write to do any good. I was shot in my right arm during the war and afterwards mashed it with a wagon bed, so it is badly used up and makes it a little hard for me to write my experience. I came to Parker county in the early days, it was unlike any part of my life, the Indians were on the reserve in Young county, and at old Camp Cooper, but our trouble with them soon commenced. A lot of us in the spring of 1859, concluded we would move them farther away, but, lo and behold when we got there Uncle Sam was there with the United States troops, and we had to back down and out, but the government moved them that fall north of Red river, and then our troubles with them began in earnest, and it continued until the buffalo hunters and McKinley got after them in 1878 and 1879, that settled Mr. Indian.

To tell of the hard times and bloody scenes during the war and afterwards, would take a mind that stored knowledge better than mine. The Indians used the scalping knife, and captured women and children; one incident I never saw in print is where the Indians on Rock creek, in the west part of Parker country, took a man's wife away from him and seventeen of them used her as they pleased, then shot an arrow in her heart, and broke it off, then scalped her alive. She lived in this condition two days and nights, long enough to tell the horrible treatment she received while in their hands. I have hunted over the ground where this occurred.

The most of the men had been in the army, or looking after the Indians, so provisions were very scarce just after the war. A bunch of women made a raid on Phelp's mill one day and took everything in the way of bread stuff on hand. I have known women to go five miles to gather wild onions to make a meal on. The clabber cheese that was eaten for bread in those days would make a man blush in our day of plenty.

To tell of the men that helped to restore order and bring peace and prosperity to this part of the country would take volumes, however, I will name a few of them: The Tackitts, Caldwells, Woods, the Harts, Pickards, Criswells, John Squires, Millsaps and Lorings.

PIONEER DAYS IN THE SOUTHWEST

They are too numerous to mention by name, but were all men of the day that didn't flinch from hardships, is the class of men that laid the foundation to make Parker county what it is today.

I had the pleasure of attending the Cox reunion west of Weatherford, Texas, in 1905, where I met a few of my old war comrades—Tom Pickard, I. C. Edwards, Allen Parker and George Johnson—also the pleasure of visiting two sisters and their families, whom I had not seen in nineteen years, and our old Company H, of the Second Texas cavalry, the first company that left Parker County, held a reunion July, 1906, in Weatherford. There I met old comrades that I hadn't seen in thirty-five years, and strange as it may seem, I knew them all but two, as I met them. But as I have digressed from my tale, I will drop back to the district court, at Weatherford, where Judge Seward was trying to hold court and Nathaniel Brammer, Edward Hopkins and the Brooks boys, with the sheriff, got on a hill northeast of Weatherford and would not let anyone come to them.

Religion was a scarce article at the close of the war, but like Elijah, there was some that had not bowed the knee to Baal, if we did have to worship under brush arbors and in log school houses with dirt floors and split log benches. In these early days men had to go armed for the redskins, so the preachers would walk to the pulpit, those brave enough to come, and lay off their brace of pistols and preach Christ and Him crucified. And what meetings we would have, and the word of God grew and prospered until that country was crowded with churches and school houses—so you see time don't go backwards but everything goes forward—the country that was once the home of the Indian and buffalo, the same is the home of the Anglo-Saxon race, and where the Indian roamed at will, the iron horse with its burden of live freight, goes dashing through the country from New York to San Francisco. Oh, what changes do come; as we grow older we see that time waits for no man.

As I went home one time from Weatherford, getting in about sundown, found my wife, baby and nephew leaving, I asked what was the trouble, my wife asked me if I didn't see some Indians just back on the hill, I told her I didn't, that she must be mistaken, she said she wasn't, she knew she saw them. I told her we would stay at home anyway. Presently our brother-in-law came, and said he saw the

33

Indians; so three families of us got our horses together and guarded them that night, and by doing so, saved our horses, but the Indians got horses all around us.

At another time, near the same place, at John Godfrey's, we had been hauling rails and I drove my wagon up in front of the door, and locked one of my horses to the hind wheel, the other one wouldn't let anyone catch him but myself. The Carnett brothers came about sundown with some good horses, tied them close to the house; just after dark the dogs commenced barking. I told them they had better look after their horses as the Indians were around, but they laughed at me, and told me my dog was always finding Indians; I told them I knew the Indians were around, but they paid no heed; so after a while I went to the door, my loose horse had come up so they got up to see about their horses, when lo and behold the Indians had slipped up and cut them loose, and were gone with them, while we had been sitting there talking. They were very much excited and one of them said if anybody would go with him he would go out, let the neighbors know, and try to get his horses back. I said that I would go with him. The first house that we came to was Jim Baker's but he wouldn't come out. The Indians had captured him the year before, and came near getting his scalp; he said the Indians were right there and he wasn't coming out. The next place was Henry Ward's; he went with us, just as we left his house we passed through a bunch of horses and struck a pony's track ahead of us. The next place we aroused the folks, got their horses and went with us, another went to let a neighbor know and the man and his son went to the bunch of horses that we had passed with a pack of hounds, and the Indians had them rounded up. When he and his hounds got there, he and his dogs together scared the red skins off. He and his boy never stopped running till they got to where we started from, but in the racket the Indians were scared away from the horses. The pony's track kept ahead of us on our round, but we never got sight of it unless it was as we returned. We came near meeting an Indian on the road but he saw us in time to give us the dodge.

The writer of this imperfect chapter, has crossed every stream in Texas, from the Canadian to the Rio Grande and crossed both of those and spent part of 1861 and 1862 at Fort Clark.

PIONEER DAYS IN THE SOUTHWEST

PIONEER DAYS IN THE SOUTHWEST

CHAPTER XIV. BY MARY A. BLACKBURN.

At one time, while I was alone, some person called "hello" at the gate and I went to the door and Charley Elms had come to tell me that the Indians were on Cowhouse Creek killing everyone. I took my four little girls and went to Mr. Blackburn's mother's about a half mile. Mother Blackburn took her little granddaughter and colored girl and we all went to William Chalk's as fast as we could run, got about half a mile away, looked around and saw fourteen Indians on their horses. They took after a bunch of horses and the last I saw of them they were going toward Poland creek. They had the two little Riggs girls up behind them at the time. Then they spied Charley Cruger, and ran him a mile and a half, to old man Nealy Robert's and then turned toward the Lampasas mountains and came upon young Plavy, ox hunting and killed him. We expected the men who had gone out stock hunting to be killed, but they came in the next day, and not seen the Indians, and found all the women and children, four families, forted up at William Chalks'. These were the Indians that killed the Riggs family.

Mr. Dallas on Dyer's creek near Georgetown, a settlement of a few houses, and one store, and my husband went down and bought two cows and calves and paid forty dollars in gold, leaving us ten dollars, but those two cows and calves were our first start in stock. My husband's father gave each of his children thirty acres of land and we built our first home on this land, where we now live, near Killeen. Thinking to better ourselves, we took a lease from Ramsey Cox, on Bear creek. This lease was nothing on earth but a wilderness full of wild beasts. Mr. Blackburn was compelled to go to mill and the nearest place was Uncle Whitefield Chalk's, about twenty-five or thirty miles distant, and during his absence I was left entirely alone with my four little girls, the oldest only about six years, and the youngest, twins of six months. At night panthers, wolves and bears all gathered around the house and were so fierce and ravenous that I was compelled to bar the door with the heavy furniture, to keep them from breaking in and eating my children, and possibly myself. For six days and nights I was left in this condition, not knowing at what minute the Indians or wild beasts might kill us. During this time Mr.

PIONEER DAYS IN THE SOUTHWEST

Blackburn was water bound and it was only after making a circuit of twice the distance that he reached home, with his meat and flour. We made one crop here, then moved back to our little farm in Nolan. On the 16th day of March, 1859, Mr. Blackburn, Mr. Elms, Mr. O'Mal and Nate Roberts, all went out that morning to look after their stock on the Lampasas river, near what is now Youngsport. Cattle were then running at large. We were living in a little log cabin, dirt floor and board chimney.

One day I had to have water and I drove my wild oxen, Tom and Dolly; they ran away with me threw me off and injured my foot. They ran on to a deep hole of water and into it they went, slide, barrel and all. On our way home from Bear creek, night overtook us on the other side of Cowhouse creek. I had one baby, and my husband was carrying the other. The one my husband carried was crying, and two panthers took after us, and ran us half a mile, and I thought every minute they would pull the babies out of our arms. When we got to the creek our faithful old dog, Cash, met us and ran them off. When we lived on the Salon our brother-in-law, Allen Rickets, and wife, came to see if we were in need of anything. My husband and brother-in-law went fishing and caught a lot of fish; ate all but the head of one large one, and next day we only had this large fish head to make soup for dinner; so fish head soup was all we had for dinner but we made merry over it and enjoyed it very much. We did not know where we would get the next meal, but that evening our good old friend, Buckskin Smith, brought us a quarter of fat venison.

I was sitting in the room one day and heard the dogs coming across the prairie, looked out and a big bear was coming straight for the house. I was so frightened I left my baby sitting on the bed and climbed the wall of the house, I left the door open, the house was made of split logs. The bear climbed, a tree within twenty steps of the door. Our good old dog, Cash, pulled it out of the tree and my husband and Mr. Smith shot it. The war broke out in 1861. My husband went out in 1864 and was gone eighteen months. I was left alone with five little girls. There were but two or three old white men in our settlement and a few negroes and I was as afraid of them as I could be. I had a very hard time; worked the corn field all day, sat up till 11 o'clock at night, carding, spinning and weaving. My oldest little girl held the candle for me while I made blankets and gloves for the soldiers, and cloth for my family. We had a pair of oxen with which I had to haul wood and water, go to mill and to church. They appointed Mr. John Roberts to kill beef for the women. I would take my oxen and wagon and go four miles for beef. I also yoked my oxen and took Mr. Blackburn's mother and my mother and family, went to the camp ground, five miles and camped. It is three miles west of Nolanville on Nolan creek. After the meeting broke up, my sister-in-law and myself came home after our oxen. We stayed at home ourselves, and thought we would have to go four miles after them, but as luck would have it, they were in the lot lying down. There were a few old preachers left on the Cowhouse creek and they brought their cradles and came over and cut my wheat for me. My oldest little girl and myself hauled it and stacked it ourselves. After a while they came and threshed it for us. I worked so hard in the corn field that year, that I never got over it. As you all know, Confederate money wasn't any account. I paid Mr. Jaconic, $75, for four yards of calico and my first baby cradle was made out of a hollow post oak tree, sawed down and split open, planks nailed in the ends and the rockers put on.

My first churn was made out of a pine box and my table out of three-foot post oak boards, my bedstead one leg holes made in the walls and poles put in, and then fastened to the one leg. I tell all the girls my piano was the spinning wheel. I have it yet. The year we lived on Bear creek we had no meat but wild meat. The old Tennessee gray squirrel immigrated from Arkansas; every bush and tree was full for two weeks, we had squirrel for breakfast, dinner and supper. They disappeared, I don't know where they went.

Women these days don't know anything about hard times and I hope I may never experience another such time as during the war.

I had three dear brothers killed in the war; J. M. Chambers and W. S. Chambers, killed in Pennsylvania under Lee, and T. H. Williams (a half brother) captured and died in Nashville. I am a faithful old Methodist, have camped on Nolan creek 58 years, at our annual camp meeting grounds. We have a beautiful camp ground, fine water and pecan groves.

We had no railroad then, took our butter and eggs to market, at Koss, Texas. 500 pounds of butter, two barrels, and 100 dozen eggs.

PIONEER DAYS IN THE SOUTHWEST

Butter was 40 cents per pound and eggs 40 cents per dozen at wholesale. Then when the railroad came to Austin, Texas, we would go there to market. I took a big whiskey barrel of butter, 500 pounds, sold all to one hotel man at 20 cents per pound.

We had one thousand head of cattle at one time, in the dry year of 1879, we drove out west to Coleman county, the winter was so severe it killed and drifted them so that we did not have many left, some drifted back home to Bell county. We increased them again in 1882; stock cattle sold for $25.00 per head. In 1887 we drove to pasture again and the winter was very severe, we lost nearly all of them again. We would brand 150 calves every year. You see we are very old Texans, my husband is 75, and I am 76 years old.

PIONEER DAYS IN THE SOUTHWEST

I was only a child of four years when my father moved from Kentucky to Texas, and my recollection of my native home is very limited. I remember the fine apple tree that stood in the front yard. My mother was rocked to sleep when a babe under the shade of that old apple tree her father planted there. The weeping willow that stood near the gate I remember well. I can see my cousin, Fanny B., in my imagination today, standing near that beautiful tree, fit emblem of her feelings, after the good-byes were said, for she was weeping too. Our old Kentucky home was within one mile of the Cumberland river, where the boats landed. My mother and I were born there, and it would no doubt have been very dear to me had I lived there until I could have realized the worth of a nice home. My father, Samuel T. Treadwell, was born in South Carolina in 1808. My mother, daughter of William and Sallie Jones, was born in Kentucky in 1816; they were married in 1833; mother died in 1872; father died in 1882. How often I have heard the old folks talk of their old Kentucky home, the conveniences, the beautiful flowers, the grand *old* orchard, the old rock spring, good houses, all went to make up the conveniences of life. They soon learned that they had left a paradise and landed in a wilderness.

We came on a boat from the old home to Shreveport, Louisiana. From there we went overland in ox wagons. We stopped at Uncle Milton Jones, my mother's brother, on the 4th of July, 1850, Independence day. I don't imagine that the people of the Lone Star state felt very independent those days. Mother was delighted to meet her brother; he had been in Texas several years; he was a Methodist preacher and a farmer, and had gone through many hardships. Away back in the old time, preachers had to work between appointments, ride horseback, swim creeks and rivers and stop wherever night had overtaken them. Many times their bed was made of their saddle blanket and their saddle bags for a pillow. Uncle Milton preached the first sermon we heard in Texas; he preached in a little log school house; it was the best the people could do and that is all the Lord requires of His people. Dear old uncle, he has long since passed to his reward and is sweetly resting from his labors. Father rented some land and made his first crop in 1851; we lived in a little log cabin that answered for a living room and kitchen; there were no luxuries for the first settlers. There were several families in the neighborhood, and they decided to build a school house. It was a pine log pen with the old fashioned stick and dirt chimney that took up almost one side of the house; the floor was of puncheons; the benches were made of pine logs split open, with holes bored in them, and round pegs, I call them, for legs, no backs to lean on for a change. I guess we as little restless tots would get tired, but kept it to ourselves, as teachers away back in the fifties were very strict in their rules in that noted school house. When finished up we had a puncheon slab that reached across one end of the house which answered for a writing desk. In place of a glass window, an opening between the pine logs was left to give light for the little pioneer children to see how to write. Miss Fannie Wood from Shelby county taught our school. I can remember but very little about her, as I was only five years old. The rude log cabin and its furnishings I remember very well. I had an adventure during that school that is still fresh in my memory.

There was a young lady going to school that had a nice book with beautiful pictures. I had a great anxiety to look at those pictures. One evening after school was dismissed and all had gone home, I decided I would go to the school house and have a good time inspecting the nice book. It was only a short distance from home, so I slipped off from my mother and I got back so quick she never missed me. Of course I went, to the school house as quick as possible. Just as I went to make a step in at the door, I saw through a big crack in the wall an awful looking beast eating the scraps the children had thrown on the hearth; it had dark brown and bright yellow rings around it from its nose to the tip of its tail. I have often thought if it hadn't been for that big crack in the wall I would have been torn to pieces by that awful looking beast; I stepped back lightly, too; and I made a run for home on double quick time. I never gave in my experience when I got back home because I knew I had done wrong in slipping away from my mother, as no doubt she had warned us children from getting too far away from the house alone. The moral of this little incident is obedience, or rather teaches obedience.

There were bear, panther, deer, turkey and a great many other kinds of game there in old Rusk county. There were no Indians to bother us. I am glad I never had any acquaintance with them, for hearing others tell of their depredations among the early settlers is horrible to think of. If there is any that deserves a monument to their memory, it is the old Indian fighters and the preachers that stood by their post on the frontier of Texas. After all the privations of the early settlers, in building homes, churches and schools, preparing to live in peace and quiet, the war cry was heard again. My experience during the civil war was very limited, as we Texas girls never heard the roar of cannon, at least I never did. I am glad I never had to witness the suffering of the wounded and dying on the battlefield. I have heard some of the old mothers tell of their experiences after the battle, dear old gray-haired mothers will soon be gone. While we call the soldier the hero of the battlefield, the mothers are the heroines. The people of Texas had many troubles and privations to endure and most of us had to work hard, raise all we ate and wore at home, and had to say good bye to fathers, brothers and sweethearts, yet Texas never suffered from the effect of the war like many other states. It was hard on us poor southern girls to have to spin and weave our own dresses; they looked awfully coarse and ugly, as many of us had never woven any cloth, much less worn such shoddy looking goods. We southern girls wasn't the kind to give up; as fast as we would get one web of our common looking cloth out of the loom we would get another ready. The women of the south were like the soldiers in the field; they thought they were fighting for their rights, and we as true southern women were fighting the home battles in adverse circumstances. We would spin, weave and sing, hurrah for the home spun dress that southern ladies wore.

The night before my oldest brother started to the army, the neighbor boys and girls met at father's and had a farewell party. Most of the young people were raised in the same neighborhood and several belonged to the same company brother did, and were going to start the next morning. It was the saddest gathering of young people I ever saw; they tried to make the evening as pleasant as possible, yet they thought of the camp life, the battlefield that was in front of them and the probability of never seeing loved ones any more was enough to make strong men weep. They would talk and sing at intervals. They sang the dear old songs in the "Sacred Harp." When the time came for the singing to close, the leaders turned to the old song, "Parting Hand."

"My Christian friends in bonds of love, Whose hearts in sweetest union join, Your friendship's like a drawing band, Yet we must take the parting hand." I don't think I joined in with the singing of the parting song, my heart was too full for utterance. The next morning the farewell words were spoken; we watched them as they rode away, and wondered if we would ever meet again, but very few of them ever came back. They left the first day of April, 1862, and on the 17th of April, 1865, my brother returned after three years hard service. Uncle Milton was at our house, leaning back smoking his old clay pipe, talking with mother; I was at my daily task, spinning and listening to the old folks talk. Uncle, looking down the road, smiled and looked at me and said, "Ermine, I see him coming." My first thought was a young man he had been teasing me about that morning. At a glance I knew him. I have no language to express my feelings; it was indeed a happy meeting after three long, weary years of waiting and watching. My younger brother, W. L. T., volunteered in March, 1863, was in Arkansas and Louisiana most of the time he was in the service; was sick a great deal of the time while in the army. The year the war closed, some time in February, he left the hospital in Harrisburg, La., on a sick furlough for home. He wasn't strong, just out of a spell of sickness, and had to travel on foot, but he was so anxious to get home that he started, against the advice of friends. He has never been heard of since the day he left the hospital. Oh, how we waited and watched for him thinking perhaps he was taken prisoner, or perhaps was taken sick and couldn't write home. But alas, after a lapse of forty-three years we have never had a word as to how or where he died. Dear, sweet brother, you are gone from me, but not forgotten. All the history that has or ever will be written concerning the civil war, there is no language that can ever picture the sorrow and broken hearts that was left in its trail. Those that were broken up, just parts of a family, homes gone to waste and many of them maimed for life. When I go to the old soldiers' reunion and see the old battle-scarred veterans, my thoughts go back to the sixties when I saw them march away from their homes and loved ones, and I weep again as I did forty-five years ago. Yes, I'm glad that my sympathy still runs out for the few that are still lingering on the shores of time waiting for the last roll call. I hope when the people of today shall read of the old veterans of the nineteenth century, that they will remember us kindly as the old tried and trusted pioneers that paved the road through the wilderness for them. When the great day shall come I hope we old fathers and mothers will be found at the right hand of God where we will have no more good-byes.

Give joy or grief, give ease or pain,
Take life or friends away,
But let me find them all again,
In that eternal day. E. K.. 1908. Born Jan. 29, 1864.

PIONEER DAYS IN THE SOUTHWEST

PIONEER DAYS IN THE SOUTHWEST

CHAPTER XVI. WAR BONNET OF THE CHEYENNES. T. J. VANTINE, QUANAH, TEXAS.

I went out in the spring of 1860 under Bill Fitzue as Captain, from McKinney, Collins County, Texas; we went by way of Jonesboro, and from there to Ft. Belknap, where we joined M. T. Johnson's regiment and then moved out about 12 miles north of Fort Belknap. The first night on guard were Ed Mires, Bill Tight and myself. Ed Mires was standing under a pecan tree, our stands were about 75 yards apart. I left my stand on some account and went to Bill Mires' stand and when I got to Ed's stand I found Bill Right there. As I sat down against the pecan tree Ed was standing under, my horse took a scare and jerked me down and then Ed Mires hallooed "Indians," and began to shoot, I got up and saw the Indian that was behind the tree from Ed and then I began to shoot with a Navy 6 and Bill Right and Ed Mires said for Lord sake don't leave us. Bill came back and the Indians shot a dozen or so arrows, and we kept firing until we exhausted our ammunition, saving only one round for future use. By that time our captain and the rest of the company came to our aid and then the Indians left us. It was so dark that we couldn't follow them. The next morning we started a scout after them. We were in a thin oak thicket and the Indians had a good hiding place in there. We found much blood next morning after the shooting, but no dead Indians. They stampeded our horses, and it took us two days to get them together. Then we moved about fifteen miles farther on towards the Brazos river and camped about two weeks there. Then we went out on a two weeks' scout and traveled about two days west, close to the Brazos river, and then we camped on a small prairie that was surrounded by tall, thin oaks. We put out our guards about seventy-five yards apart and along about 10 or 11 o'clock at night the Indians began to show up in five or six different places around our camp. The boys began to shoot, four or five at a time. They kept it up all night. That was the first trip for some of the boys, and they were all excited. There were about seventyfive Texas rangers in that scout, who wanted to move the camp that night. The others did not want to go because they were afraid the Indians would molest us. If we stayed till morning we would have a chance to get out, but the next morning the captain said we had better go back to the regiment. There wasn't enough of us to *fight* them then, and when we got back to camp we were ordered out on another scout at the head of the Washita river. A large number of Indians were headed for the Wichitas. We at once went after them but didn't strike their trail. We were out ten days before we turned back to camp. Then we moved camp and started for the Wichita mountains, and camped on the west prong of Otter creek at Colonel Van Daran's old camp. We scouted that country about two months and killed buffalo and antelope and hunted Indians for pastime. The Indians were reported to be there and threatening our camp. We had been hunting for them about a week and couldn't find them. Dave Wash and I were out hunting bear one day and we found a mountain where the rocks were broken open and made an opening where one could walk in about seventy yards, and there was a bear den in there and we went back to camp and our colonel, M. T. Johnson, gave us orders for no one to leave camp without orders, but I wanted to kill a bear by myself, so the next morning about the break of day I slipped out through the guards and struck out for the bear den about four miles away. I got about three miles but there were many coyotes and lobo wolves howling. I heard one that howled different from the rest of them. I began to hunt to see what it was, and I saw a big Indian standing on the bank of East Otter creek. I drew my gun down to fire. I thought that I could hit him, but I missed, the ball striking right at the left of where he stood. I had a muzzle loading Mississippi rifle which threw an ounce ball. I went through the motion of loading right quick and threw my gun down on him again, and he ran into the brush. I turned then and ran behind a hog back mountain and ran about half a mile to where I left my horse. I had gained the spur of the mountain when I looked to see if the Indians were coming on the other side, but I saw none. When I looked back I struck my foot against a rock, but I lost no time in the fall. I ran against my lariat pin and knocked it out and done my rope up as I ran and jumped on my horse and went back to camp. I didn't let any grass grow under his feet. When I got to camp my captain came out and took hold of my horse and asked me what was the matter. I told him I shot at an Indian up on East Otter creek. He said I might consider myself under arrest and he took me to the colonel's headquarters and they assessed my fine at ten days on guard duty, two

47

hours on, and four hours off. I stood two turns. On the third turn I called to the corporal of the guard that my time was up, and he started around with the relief, and before he got to me I laid down and went to sleep, and the fellow that was going to stand where I stood, begged them to let me lay there for company, and so they assessed me ten days more, two hours on and four hours off. They started a scout after the Indians the day I came in, and found fifty of them close to where I shot at that one. They ran them up about Fort Cobb and the Indians all scattered and they couldn't follow them any farther. Then we moved camp about fifteen or twenty miles on the south prong of Red river, and that night they put me on guard, right close to a big slew, and the guards were about seventy-five or eighty yards apart. About 10 o'clock that night a big bear came splashing through the water. He passed a man by the name of Vanvaris and then by me, and then by a fellow by the name of Van Winkel. The bear had caught a mussel and was sitting upon his haunches eating it. He was a fair target for Van Winkle, and he fell dead right there. Van Winkle hallooed "Indians," and ran into camp, but I didn't believe it was and stayed at my station. When Van Winkle ran into camp the whole company came out. They asked me where the Indians were, and I told them I hadn't seen any, and didn't think there were any Indians here. Then they asked me where Van Winkle stood, and I went down and showed them where he stood. Then I looked over the weeds and water and saw the bear lying where he had given him a dead shot, and they went back to camp. That was the last night of my sentry duty. We scouted around there about two weeks longer and then we started out on the famous scout. We crossed the South Canadian where that emigrant train was captured by the Indians in 1849, and the people all murdered; it was not very far from the old Adobe fort. We went on and crossed the North Canadian, then we struck across the plains to the Cimarron river. When we got to the river we traveled its banks four days and a half. We ran out of provisions the day before we struck the Cimarron; then we traveled up the river four days and a half, then we lay idle a half day. Every Indian camp we struck the ashes was full of beads.

A man by the name of John Huff and I were out hunting; we both had big guns; he claimed he had the best gun and I claimed I had the best. We were standing on the side of the mountain and there was a stump on the other side of the valley on a mountain, we shot at the stump to see who had the best gun. John Huff shot first and he hit right at the root on the right hand side; I shot and hit right at the root on the left hand side; then we went over to the stump and called it a tie shot. While we were up there looking around there was in a big crevice in a rock and an old dead Indian lay in there wrapped in a blue blanket, and I wanted Huff to go down and get him but he wouldn't do it, so I told him that I would go down; that I wasn't afraid of a dead Indian, so I went down and unwrapped him. There was nothing there but the bones and hide, so we just took a little bone off the shoulder and left the rest lay. We went back to camp and told the boys and a good many of them went and got a bone to take home with them. One Indian guide said this Indian was a brother of a Chickasaw Chief. After that we started on another scout, four of us. We went up into the mountains and saw a Mexican lion laying upon a rock jutting over four or five feet, and we took a shot at him, and he came rolling down nearly to where we were before he stopped and he showed fight at sight of us and nearly scared our horses. We couldn't get within fifty yards of him. We shot him several times, but he didn't die, and we were afraid to go to him on foot. We then took our scout about half a mile further and saw three Indians. We were afraid to follow them on account of running into a big bunch of them, and also getting cut off from the camp; so we went back to camp. There were fifteen men out that never came back and they fared worse than we did. They traveled all day and didn't get anything to eat. They killed a wild cat and ate it and the next day after they eat the wild cat they killed an Indian pony. The Indian had just gotten off of him and went to the brush. After they killed him they cut out of him what they wanted and went on about five miles and stopped and got supper. They went on again until away after dark so as to dodge the Indians; that was the way we all traveled, so we would not have any fire where we camped. Those men traveled straight for the Wichita mountains. Before they got to the mountains they ran onto part of their own regiment and took them to be Indians, and their regiment also took them to be Indians, and they all began to shoot and broke for the brush. When they got together they were surprised but there was great rejoicing. They all went to the mountains with the regiment; they got through two days before we did. Now I will refer back to our camp on the Cimarron river.

We started the next morning traveling east up a big level flat. There was a rock about ten miles ahead of us that was fifteen feet high and as big as a house. It was flat on top, and was called the Indian rock. When the Indians were traveling through the country and came to that rock they all left something—such as rings, earrings, and beads. It seemed as though they worshipped that rock. After we got there our Indian guide didn't know the road any farther, and so we traveled through an unknown country without any guide for four hundred miles. We traveled that day about ten miles farther and camped for the night with but little to eat or drink. Next morning we started about sunrise and traveled about five miles and came across an Indian trail going east. Then we went ten miles farther and camped. We had killed three deer that day and some rabbits, so we fared pretty well for supper and breakfast. We started pretty early next morning and traveled about fifteen miles and struck the brakes, and there we struck another Indian trail about the same size the first one, going the same direction. Traveling on about five miles farther we struck a stream which we took to be the North Canadian river; there we saw quite a few Indian signs and traveled on about six miles and camped for the night; then we went through a smooth prairie country for about fifteen miles. Our hunters had killed some deer that day so we had plenty to eat. That day we traveled about ten miles farther. That evening as the hunters came in they saw three Indians right south of us; that was all the Indians we saw that day. We camped there that night but was off next morning by daylight. We went ten miles, and crossed a stream that we took to be the South Canadian. We traveled all that day and didn't have much to eat that night. The next day we traveled through a rough country and killed a bear and two deer and one or two turkeys. We ate the deer meat for breakfast and the bear meat for supper and camped for the night again. We started about two o'clock in the morning and traveled over a pretty level country for ten miles. We stopped then to let our horses graze and eat breakfast; it was about eight o'clock when we ate. We started on the march again about ten o'clock; we traveled till about two o'clock and then we let our horses graze an hour or so; some of them had little to eat and most of them didn't have anything. We traveled on about seven miles farther, when there was a bunch of about one hundred head of wild horses ran through our ranks. As we were marching there was a colt that joined us from the bunch of wild horses, land we let him follow along with us. We traveled about five miles and came across a lone buffalo: they killed and skinned him and took his hide for moccasins and bridle reins.

We went down to the creek and camped, there was plenty of wild grapes and nice running water. Three men and myself went a quarter of a mile below the rest of them and got our suppers. Later the three men saddled their horses and rode off, leaving me alone. About ten o'clock at night I thought I heard the Indians hello right close to me. I took my horse and moved him about a hundred yards farther up the creek, closer to where the company was camped. I ran onto a man in a ditch asleep. I woke him up and told him I heard Indians, and that we would have to stand guard, and he agreed to stand guard two hours off and two hours on till daylight. I stood the first two hours and woke him up and he was to stand two hours. In about an hour I woke up and he was asleep, so we both stood guard from then until morning. About four o'clock in the morning those three men who left me alone came back and camped right close to where they had eaten their suppers that night, camping by an old dry stump about twenty feet high; it was covered with old dry vines and they built a fire by the stump and the vines caught fire and made a big light, they could see it for miles and we thought it was Indians, and we were as still as death for about five minutes, till the fire went down and we hailed them before we went up to the fire, and when we went up we saw that they were the men who left me that night, and they asked me if I hadn't followed them all night. I told them no, that they had traveled all night and had come back to where they had eaten their supper's the night before. I told them I would show them when daylight came where they had eaten their suppers. We ate our breakfast, then about sunrise and started on our journey east, right down the creek that we were camped on; traveled about three miles, rode up on a high rocky mountain to look around and see if we could see any Indians or the rest of the company coming, but we didn't see any Indians anywhere; then we looked west and saw our company coming. We waited till they were within four hundred yards of us to see if they were Indians or our boys. We saw they were our boys and we came down to fall in line with the company. The three men I camped with the night before were aiming to leave the company and go to Fort Belknap by themselves, but got lost and came back to where they started from that night. We traveled on till about eleven o'clock that day and stopped to graze our horses a couple of hours when our hunters came in. They reported seeing five Indians about a half a mile south. All we had for dinner was Mesquite beans and hack-berries and a few prickly pear apples. We started again on the march about one or two o'clock and traveled about ten miles. We came to what they called Dog Town or Red river, about half of us camped on one side and about half on the other side. We had nothing to eat or drink when we camped. We dug down about three feet in the river and found a little alkali salt water. The party that camped on the west side of the river then sent us word that they had killed three buffaloes and found water. We all went over there and camped with them. We had plenty meat to eat and plenty of water to drink; we laid down and all went to sleep, and our officers were all out and we didn't put out any guards. About four o'clock in the morning, just as the moon was going down we heard two shots, and then the Indians began to yell and ran through our camp, taking sixtytwo head of our riding ponies and pack mules. Captain Sull and Pete Ross fired at the Indians as they passed right over them. I heard the captain say to shoot at the yelling Indians and to shoot downward. The yelling and shooting scared me so I couldn't keep my hat on my head. There was a pond of water close, and two or three of our boys were scared so bad that they ran and jumped into that pond of water. I had turned my horse loose that night and just left him drag his lariat and I had to go and hunt him up myself, because I wouldn't ask anybody else to go with me. I went up the river about six hundred yards with my army six in my hand. I found my horse, got on him and looked around and saw another horse. I went and got him and took him to camp with me. The owner of the horse came to me when I got to camp and said he wouldn't have went out after him for a dozen horses. When daylight came our colonel called for all of those who had good horses to follow the Indians, and I was one of the men that went with the detail after the Indians. We went about six miles when we found the Indian Chiefs head dress. It was fine polished buffalo horns and covered with velvet and painted feathers and beads; then we went on about half a mile farther and came to a hill. We sent four or five men to look over the hill and saw our horses; the streams forked there. These were two tribes of Indians; one tribe camped on one side, and one on the other. They were herding our horses between the two tribes, and there were so many of them we didn't tackle them. We went back to camp and the orders were that we wouldn't travel any that day. Some of the boys who lost their horses cried. We stayed there all day, and along about three or four o'clock in the afternoon there were a few shots fired through our camp. We had lost fifteen men on the Cimarron and Major Fitzue thought it was them shooting through our camp, thinking we were Indians. So Major Fitzue went out about five hundred yards and hallooed at them and waved his hat, and he said they shot so close to him that he knew it wasn't our boys. Then we stayed there until dark and piled our saddles and pack saddles and everything we couldn't carry with us and burned them up. We threw our cooking utensils into a hole of water. We traveled right down the bed of the river ten miles and went up in the sand hills about three miles and camped. Most of us had no water or anything to eat that night. The next morning we traveled till about 11 o'clock and got off and left our horses graze and the men on foot have a rest, and we had nothing to eat or drink. We started again and traveled two miles and left Joe and John England under a mesquite bush, played out. We traveled about a mile further and left Frank Hunter under another bush. I was carrying a man behind me and I stopped to get him on my horse, and the man that was on behind me objected and hallooed at Major Fitzue that I was going to get off and make him get off and take Frank Hunter on. Major Fitzue said to tell me to get on and come on or he would have me dismounted in half a minute. I told Frank if we found water anywhere close I would come back and get him. We went about six miles and found water, but it was so salty we could not drink it. The officer was quarreling down in the bed of the river, and there were about twenty-five men who got down along the side of the banks and began to pray. Word came down the river that there was plenty of nice water about four hundred yards above. We all started up the river to the water, and when we got there it was alkali or jip, and they drank so much of it that it made them all sick. The man who was riding behind me jumped off into the water, clothes and all. There was a man who came down the river and told us that about a quarter of a mile above was a nice stream of water and a good nice fortification. We all went up there and found it all O. K. and then sent back and got the boys we left and sent out hunting parties. My mess didn't kill any game. I was at Captain Burlson's company, and his company had killed three deer, and he gave me half a deer to take to Major Fitzue, and told me to tell Major Fitzue to divide it among his men that had nothing to eat. I had two shirts on, so I cut off all the flesh I could and poked it in between my shirts, and told the major that Captain Burlson told him to divide it among the men who had nothing to eat. The major told me to cut me a piece and say nothing, so I cut it in two and went to my mess and told them how I treated the major, and they said it was all right, that the major was always a rascal, so we ate our supper and when night came we all went on guard. We carried rocks and logs and made a fort. They put about fifty men in about two hundred feet, and none of us slept that night. Our lieutenant-colonel, Smith, went up and down the lines and talked to every man. He told them they would have to stand and fight as though they knew no danger. Then the guard sent word down the river that there was a large force of Indians coming down the river, and Colonel Smith kept going up and down the lines telling his men not to shoot until they could kill an Indian. He told them if they ran they were gone, but if they stood they had a good chance to save themselves. The Indians came up on the other side of the river from us, and the moon was shining as bright as day, when all at once the Indians made the awfulest whistle I ever heard. I thought sure they were coming, so I braced up to face the storm. They turned back, and Smith told us to hold our places, for they might come from some other direction, but they never came in that night; we stayed there and didn't sleep any that night. The next morning we found that about half a dozen men had their horses saddled to get away, while we did the fighting.

We got our breakfast and started out about sunrise and traveled about ten or twelve miles and found a pretty good place and thought we would stay all night, and sent out hunting parties to get game to strengthen the boys up a little. We got enough game for our supper and breakfast. We had to travel slow on account of our men on foot. We started the next morning after breakfast and traveled about twelve miles farther. That day our south guard saw Indians about half a mile away, but as there were only two, we paid not the least attention to them. We traveled until we came to a good camping place where we had plenty of water, but we had very little to eat, enough for supper but nothing for breakfast. We traveled on ten or twelve miles again, and our men on foot got so they could walk better than they did all forenoon. They picked mesquites, beans and prickly-pear apples as they went along for their breakfast. Nothing more of note happened for four or five days. When we got to the river we got some buffalo and steers that had strayed off from the settlers. Before we got there we killed three or four bears in one place, and up in the shinoaks we saw some musk hogs. When we got to the river we found out where we were. Some of the boys had been there before. We crossed Peas river about twenty miles above Mule creek, where the Indians captured Cinthy Ann Parker. After that there was nothing more of note happened until we got to the Wichita river, then we camped on the Big Wichita a day or two, and the colonel sent four men to Fort Belknap to get provisions, about sixty miles, and I was one that went. We traveled about thirty miles and met Till Yelton and others about midnight with a train of pack mules and guards, with provisions for our scout. We were gone so long that they were going to hunt us up, to see what had become of us. A fellow by the name of Jim Lewis, had eaten so much bacon and stuff that it made him sick, and we had to roll him all night, so we didn't get to Fort Belknap till about noon the next day. We stayed at Belknap about four or five days and there were about two hundred of our men who came in, and we were discharged. John E. Bailor went out with four or five men the day before I was discharged and he came in the next evening with seven Indian's scalps, and that was the last of the scout. I will go back and tell of some more Indian and buffalo chases that I wasn't in, but some of my company were.

Four men up in the Wichita mountains upon seeing a like number of Indians gave chase. They took after them and the Indians broke for a creek where it was awful brushy. They got within about three hundred yards of them, but couldn't get any closer, so they ran them for about three miles till the Indians got into the brush and there they gave up. White men never follow them any farther than where they strike the brush, then they quit them. When the men came back they said an Indian could get more ride out of a horse than a white man could. Every time our boys went to shoot, the Indians would gain ground on them, but they had an interesting chase of it. And a few days after that there were five of us went out on a buffalo hunt. We got after some buffaloes and wounded two or three of them, and one of the wounded buffaloes stopped and showed fight. He made a lunge at one of the horses and hooked one of them in the side and knocked him down and ran over him, the man was kicked about fifteen feet. He got up, ran and jumped into a gulley and said: "Wasn't I lucky not to get killed?" He then jumped on behind another man and went back to camp. Two months after that Phillip Yelton went out buffalo hunting by himself. He saw some buffalo about two hundred yards from him, so he got off his horse and slipped up on one and shot at him. He thought he had killed it but when he came close to him he jumped up and made at Phillip. He shot at him again and the buffalo whirled around six or eight times, and every time he whirled around he would shoot him again. Three or four of the other men saw the fight, and got there in time to attract the buffalo's attention so Phillip could get away. After he had emptied both of his six shooters, the men ran up and patted him on the shoulders and said: "Bully for you, Phillip." This closed our campaign for the year 1860.

THE END.